ıa Design

Small Antenna Design

by Douglas B. Miron, Ph.D.

AMSTERDAM • BOSTON • HEIDELBERG • LONDON
NEW YORK • OXFORD • PARIS • SAN DIEGO
SAN FRANCISCO • SINGAPORE • SYDNEY • TOKYO

Newnes is an imprint of Elsevier

Newnes is an imprint of Elsevier
200 Wheeler Road, Burlington, MA 01803, USA
Linacre House, Jordan Hill, Oxford OX2 8DP, UK

Recognizing the importance of preserving what has been written, Elsevier prints its books on acid-free paper whenever possible.

Library of Congress Cataloging-in-Publication Data

(Application submitted.)

British Library Cataloguing-in-Publication Data
A catalogue record for this book is available from the British Library.

ISBN-13: 978-0-7506-7861-2
ISBN-10: 0-7506-7861-5

For information on all Newnes publications
visit our website at www.newnespress.com

06 07 08 09 10 10 9 8 7 6 5 4 3 2 1

Printed in the United States of America.

I dedicate this work to my mother,

Florence Ethel Coolidge.

She gave me love, liberty, and a backbone.

Contents

Preface

Miniaturization of electronic systems has accelerated over the last few decades and this process feeds expectations for even smaller components and systems in each new generation of equipment. Antennas have not been exempt from this pressure to be made smaller. Often, the result has been the use of antennas that are reduced in size without regard to their performance. This has led to needlessly poor system efficiency and reduced range. In this text we discuss the limitations on small antenna performance, possible trade-offs, recent developments, detailed design and optimization.

Antenna performance is fundamentally a function of size measured in wavelengths at the operating frequency. "Electrically small" antennas are those that are small compared to the wavelength, not necessarily small compared to the people who use them. The wavelength at the middle of the AM broadcast band is 300 m, so a tower 30 m tall is called electrically small even though it's 15 times the height of a tall man. Small in the antenna business can mean "electrically small," "low profile," or "physically small." Historical applications have included mine communications, broadcast transmission and reception, and mobile radio communication for both military and civilian uses. Present and future applications include the historical ones plus mobile telephones and handheld combinations of telephones and wireless data links for video and computer mobile networks, and wireless data networks that include both stationary and mobile elements. Versions of these networks are being designed and deployed in the obvious area of personal communications, and areas as diverse as medical monitoring and industrial production. The performance and efficiency (battery life, for example) of any of these systems depend in a very basic way on each device's ability to get its signal out and capture the signals from the other elements in the network. The antenna is the component that does it.

The intended reader for this book is the design engineer with a B.S.E.E. degree. Chapter-end problems have been written so that the book can also be used in a senior-level elective course. Most people graduating from such a program have an exposure to electromagnetic theory but no experience in RF circuits or actual antennas. Some mathematics is necessary to understand the concepts presented, but few derivations are given. Each topic will begin with a discussion of the physical principles involved. The simplest possible illustration will be used first, one with analytic results available if possible. Then more complex and more practical versions of the topic will be presented. The emphasis is on the intelligent use of formulas where available and applicable, and numerical modeling. The analytic results for simple structures not only provide useful guidance in themselves, but they also serve as checks or standards for the numerical modeling process. The ideal situation is achieved when analysis, numerical simulation, and experimental results all agree, and published results are used whenever possible.

A major concern of this work is to bring small antenna design into the current computational environment. Familiarity with the Windows operating systems, C++, and MATLAB® is assumed, but not entirely essential. A reader familiar with C will not find it difficult to read the program listings in the book. Some historically important data has been converted from graphical form to curve-fit equations so that the entire design process can be done on the computer. Many original modeling programs have been written for both traditional and novel antennas and supporting structures.

Teaching is the third way of learning. First, one learns as a student, then as a practitioner. To teach, one has to understand even better to pass the art and science on to another. Unlike this Preface, the book is written mostly in the style of informal lecture and conversation. It is as much a story as a textbook.

About the Author

Doug Miron earned his degrees from Yale and the University of Connecticut. He was educated as a general-purpose EE, later specializing in control systems. He has worked, taught, and published in nearly every area of electrical engineering. He began his professional interest in radio frequency systems, circuits, and antennas at Hermes Electronics Ltd, Dartmouth, Nova Scotia, in 1974. He continued research on RF circuits and small antennas while teaching at South Dakota State University from 1979–1996. Among other activities, he has published several articles and papers on numerical impedance matching, LC and microstrip circuits, numerical antenna simulation, and also discovered the volume-loaded small dipole.

What's on the CD-ROM?

Numerical methods are the main tool used in this book. The main numerical tool is NEC2, the U.S. Government-developed code for modeling wire antennas. The folder NEC has source code and executables for the current user-modified version of this code. The root folder also has a public-domain GUI in a .zip file. In addition, the NEC folder has many C++ programs to generate NEC input files for various antennas and structures. These programs read text files with numbers describing the geometry and operating conditions for each antenna and structure. Examples of such text files and other utility programs are also in the NEC folder.

MATLAB was used for data analysis in many of the examples in the book. Utility programs to find the equivalent circuit Q, bandwidth, resistance components, and antenna voltage for a given power input are in the MATRF folder. This folder also contains programs for impedance-matching design, both narrow-band and wide-band. Also, there are programs for generating data used in some of the book's examples, including the NEC basis functions and curve-fitting with these functions.

CHAPTER 1

Introduction

1.1 What Is Small?

"Small" is obviously a comparative term, and so one must ask what is the reference standard. For ordinary usage, we may use as reference the middle of the size range of objects in the class being discussed. For example, among dogs a dachshund is small and a St. Bernard is large. But the medium dog is small compared to the medium human, whether we're talking about height or weight. For antennas, it's hard to say what the middle size is, so people talk ordinarily in reference to the human scale. I call this *physical size*. That is, an antenna that fits in your hand is physically small, while one that is 20 m tall is physically large. Physical size, along with the environment in which the antenna will be used, is very important in the mechanical design of an antenna, but it is only secondary to the electrical design process.

The scale of interest for electrical design is the free-space wavelength at the operating frequency. From physics, we know

$$\lambda = \frac{c}{f} \tag{1.1}$$

where c = speed of light, f = frequency, and λ = wavelength. Generally speaking, an antenna is considered *electrically small* if its largest dimension is at or under $\lambda/10$. If f is in MHz, it is convenient to use c = 300 Mm/s, because then λ is in m. At the middle of the AM broadcast band, f = 1 MHz, so λ = 300 m. An antenna that is $\lambda/20$ = 15 m is electrically small but physically large. At the middle of the FM broadcast band, f = 100 MHz, so λ = 3 m. A $\lambda/20$ antenna will be 0.15 m long, which is physically small. At f = 2.4 GHz, a cell phone band, λ = 0.125 m and a $\lambda/20$ antenna would be 6.25 mm long, physically tiny. For all three of these applications, the electrical design considerations for a given antenna type are the same.

A kind of hybrid size category is *low profile*. This usually means the antenna is short compared to the object on which it is mounted, and usually turns out to be electrically short as well, but not necessarily electrically small because its width dimensions can be λ/4 or larger. Typical applications include vehicles, especially military ones, and handheld radios and phones.

Since I am concerned only with the electrical design aspect, I present material on electrically small and low-profile antennas in this text. Larger antennas are extensively covered in numerous books on general electromagnetics and antennas. In sections 1.3 and 1.4, I describe briefly some of the past, present, and possible future types you might see. They, along with others, will be described and analyzed in detail in later chapters.

1.2 What Are the Problems?

The nature of antennas, definitions of performance, and effects of size on performance are described in detail in the next two chapters, but I think a few introductory words here will help the reader to understand the next two sections a little better. An antenna is a device whose purpose is to convert between circuit power, voltage and current at the radio terminals, and radiated power carried in an electromagnetic wave. Without a size constraint, most antennas would be built close to size multiples of λ/4. This is because the terminal impedance of the antenna for this condition is real and easily made compatible with the radio or transmission line to which it is connected. Antennas sized this way are called *resonant-length* or resonant-size. When traditional antennas are operated at frequencies for which they are electrically small, their input impedances become more and more reactive and this makes it harder to transfer power between them and the radio. Also, the coupling the antenna provides between the circuit terminals and the wave becomes less, whether we're talking about transmitting (generating a wave) or receiving (extracting power from a wave). For a series model of the antenna impedance, the reduction in coupling is manifested in a reduction in *radiation resistance*. This makes circuit loss and antenna copper loss relatively more important in degrading system efficiency.

1.3 Some Historical Small Antenna Types and Applications

The antennas we see around us every day show a variety of shapes. Most people are familiar with the large curved reflectors used in radar and satellite receiving systems. An observant person will notice that often there is a short rectangular or circular pipe with a flared opening, called a horn, in front of the big dish. Both the dish and the horn are electrically large. At the 4-GHz satellite TV frequency, a 3-m dish is 40λ in diameter. Its purpose is to scoop energy out of the passing electromagnetic wave and focus it into the horn. I mention this because it provides a picture of the idea of an antenna having an *effective area*. Almost all the other antennas you are likely to see are wire or wire-equivalent structures. The typical set-top TV broadcast antenna is a two-wire "rabbit-ears" which should be adjusted to be λ/2 long for the channel being watched. This antenna is for the VHF channels. Sometimes the set-top antenna also includes a small loop to be used for UHF reception. This loop is typically electrically small.

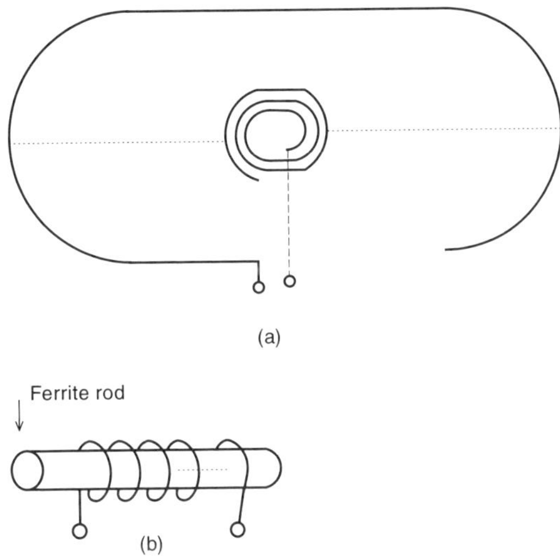

Figure 1.1: AM receiving antennas. (a) Flat coil, typically 25 by 20 cm, with many turns of insulated thin wire. (b) Ferrite loopstick. Anywhere from 3 to 30 cm long, with many turns of enameled fine wire.

AM radios in the '40s and '50s commonly had a flat coil of many turns attached to the inside of the back cover. As radios became smaller in the '60s, these coils were replaced with solenoidal coils wound on ferrite cores. Figure 1.1 illustrates these antennas. From the discussion in section 1.1, you can estimate how extremely electrically small these antennas are. They function by coupling to the magnetic field component of the passing wave, which induces enough voltage in the coil to produce a signal larger than the electrical noise generated by the receiver itself. A tuning capacitor is usually placed in parallel with the coil. The two together form a resonant circuit to limit the bandwidth at the input of the first stage of the receiver. If an open-wire antenna were used, a separate coil would be needed for this purpose.

Until the ’70s, all small antennas were variations on coils or open-wire designs. Formally, open-wire antennas are called *dipoles* if they have two wires like the rabbit-ears, or *monopoles* if they have one wire. When transmitting, dipoles are driven by a voltage applied across a small gap between the two wires. Monopoles are driven by a voltage applied between the wire and a ground system. In the case of AM transmitters, the wire is actually a steel tower whose height is $\lambda/4$ at the operating frequency, which is why they have aircraft warning lights on top. At VLF, 30–300 kHz, and ELF, below 30 kHz, $\lambda/4$ is an impractical height. There are transmitters operating in these frequency ranges for navigation beacons and submarine communications. For these applications, the vertical wire is an electrically short tower with a layer of horizontal cables going from the top of the vertical wire (tower) to supporting poles. This arrangement makes the antenna somewhat like a capacitor that happens to radiate. This method of improving the performance of a short vertical is called *top-loading*. Since it increases the capacitance over what it would be for the short tower alone, it decreases the terminal reactance, which makes the power transfer problem easier. It also raises the radiation resistance. Most people will never see a VLF transmitter, but the same principle is used by radio amateurs, especially in the 150, 75, and 40 m bands. So you have a reasonable probability of seeing a top-loaded monopole in someone’s back yard.

The ground system for monopole applications mentioned above is usually an artificial ground made by laying either a metal mesh or a number of radial wires over as much area as possible. Another monopole that one sees more frequently than the top-loaded version is the whip mounted on a vehicle or a handheld radio or radio-telephone. In the case of the vehicle, the ground is the metal body. In the case of the handheld radio, the ground is the metal enclosure, inside the plastic case, of the radio box. The performance of electrically short whips can be improved by including a series coil, either at the antenna base or part-way up. This is called *coil-loading* when the coil is part-way up, and has better aerodynamics than top loading. Coil-loading is not as efficient as top-loading, but it does provide a real input impedance and improves the radiation resistance. Sometimes top-loading and coil-loading are combined to get an electrically short antenna with a real input impedance [1]. This arrangement is sketched in Figure 1.2.

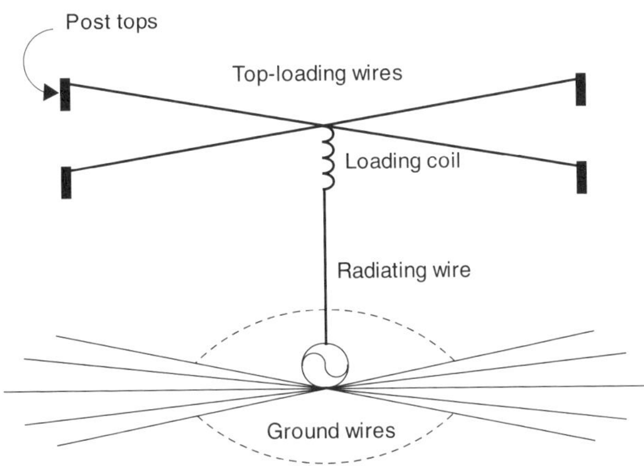

Figure 1.2: Sketch of a multiply-loaded and tuned monopole for the HF band. The ground wires are laid on the earth or buried. Four top radials are shown, but performance can be significantly improved by using more.

1.4 Some Present and Future Small Antennas

Numerically, the most common present and future potential application of small antennas is in handheld radio devices such as cordless and cellular telephones. Very recently, these phones have incorporated video and data, and radio-linked computer networks are being deployed in small numbers and selected urban markets. While these may not be all the future applications, the requirements for future applications will be pretty much the same as for the present range of uses. The frequencies used for these applications have historically been as low as 49 MHz, but as of the '80s, the lowest frequency is 800 MHz. Many systems today are using 2.4 and 5 GHz. As can be seen from the first section, electrically medium-sized antennas are physically small, but still big enough that they aren't inside the device package. The present and future utility for electrically small and low-profile antennas is therefore in the fact that they can be hidden in the packaging of the radio.

The biggest change in single-antenna designs of the last 30 years is the advent of printed-circuit antennas. Most are low-profile types [2], but there have been some that are electrically small and more are being developed. Indeed, new designs and improvements are published nearly every month. The earliest and most common microstrip antenna is the rectangular printed patch. This is effectively a wide piece of microstrip transmission line approximately $\lambda_g/2$ long, which makes it resonant at the operating frequency. λ_g is the wavelength in the dielectric material of the printed-circuit (pc) board. The radiation originates mainly from the open edges at the two ends of the transmission-line patch. The radiation field and input impedance depend strongly on the resonance, and the resonance is quite sharp (narrow frequency range) so that a major effort in improving the bandwidth of patch antennas continues to this day. Recent designs involve multiple layers of alternating dielectric and copper, many different shapes of both patches and open copper patterns, and many different ways to apply the signal. There are several books currently available covering printed antennas, so they are not discussed further in this text.

Research has continued in other approaches to solve the problems of small antennas. One approach that has received attention since 1960 is coating a wire with a layer of dielectric or magnetic material, or both. While the term used sometimes is "coating" [3], the actual geometry frequently has more material than wire. These antennas are not discussed further in this text, because they have shown little improvement over bare metal designs, for practical coating materials.

A basic idea about small antennas, put forward by Wheeler [4], is that a small antenna will work better if it makes better use of the volume it occupies. To this end, a study of dipole shapes led me to the principle of volume loading [5]. The objective is to achieve impedance resonance in the structure itself, and this is accomplished by using a thick body as the end-loads for a dipole. This is illustrated in Figure 1.3, which shows a design in which one radiating element is coiled, and the end-loading bodies are frameworks that occupy about 20% of the height of the antenna. This approach maximizes the capacitance in the space available, and permits resonance with practical-sized wire or tubing in the radiator. Another recently published approach is called the folded spherical helix [6], illustrated in Figure 1.4. The multiple coil arms are all grounded except for the one with the little circle showing the place where the source voltage is applied. The structure has both internal inductance and capacitance and a number of resonant frequencies at which it is electrically small. The author designs it to operate at the first series resonance, at which it has an excellent impedance match. Neither of these antenna types has yet found commercial application, but they are very promising.

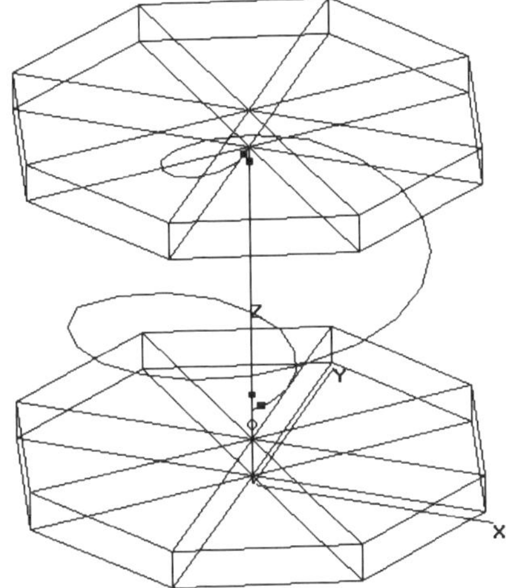

Figure 1.3: A volume-loaded dipole. This example has two radiating wires, a coil and a straight vertical wire. The squares in the drawing represent relay contacts to switch between the radiating elements.

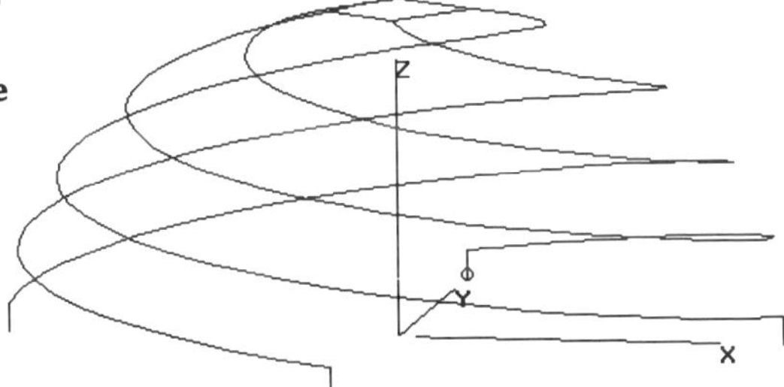

Figure 1.4: A folded spherical helix antenna. There are four arms, three of which connect to a ground plane. The circle on the fourth arm represents the applied voltage source.

References

[1] C. J. Michaels, W7XC, "Evolution of the Short Top-Loaded Vertical" in *Vertical Antenna Classics*, ARRL, Newington, CT , 2001, pp. 86–90.

[2] R. E. Munson, "Conformal Microstrip Antennas and Microstrip Phased Arrays," IEEE Trans. on Antennas and Propagation, Vol. AP-22, no. 1, January, 1974, pp. 74–78.

[3] J. R. James and A. Henderson, "Electrically short monopole antennas with dielectric or ferrite coatings," Proc. IEE, Vol. 125, no. 9, pp. 793–803, Sept. 1978.

[4] H. A. Wheeler, "Fundamental Limitations of Small Antennas," Proc. IRE, Vol. 36, no. 12, pp. 1479–1484, September, 1947.

[5] D. B. Miron, "Volume Loading—A New Principle for Small Antennas," ACES Journal, Vol. 14, no. 2, July 1999.

[6] S. R. Best, "The Radiation Properties of Electrically Small Folded Spherical Helix Antennas," IEEE Trans. on Antennas and Propagation, Vol. 52, no. 4, pp. 953–960, April, 2004.

CHAPTER 2

Antenna Fundamentals I

2.1 Electromagnetic Waves

Before 1800, we knew from Coulomb's experiments in France that static electric charges obeyed a force law similar to Newton's law for gravity, and that magnetic materials seemed to be made of magnetic charge-pairs that also obeyed such a force law. By 1830, Oersted in Denmark had observed that an electric current produced a magnetic force, Ampère in France had quantified its law, and Faraday in England had established the law that a changing magnetic field produces an electric force field. J. C. Maxwell in England put these facts together to conclude that time-varying magnetic and electric force fields generate each other (you can't have one without the other), make waves, and that light is electromagnetic waves. His theory was published in 1873, and in 1888 H. Hertz in Germany demonstrated by experiment the existence of radio waves. Wireless (radio) communication was on its way.

For historical reasons, the symbols used for electromagnetic field quantities have odd names and relations that are not quite intuitive. Two of the symbols, E and B, are force field vectors, but E is called electric field intensity and B is called magnetic flux density. For the other two, D is called the electric displacement vector and H is called the magnetic field intensity vector. We now have a better understanding of what these vectors represent, and so I call E the *electric force field vector*, B the *magnetic force field vector*, D is still the *electric displacement field vector*, and H is the *magnetic displacement field vector*. E is the force per unit charge at a specified point in space-time, and B is the force per unit current-length at a point in space-time. D is a field quantity that represents the fact that somewhere charge has been separated from its natural neutral condition. Its relation to the charge distribution is independent of the material medium. Likewise, H is a field vector that represents the fact that somewhere an electric current is isolated out of the normal random charge flow in a material medium. There are no magnetic charges; magnetic fields are produced by the motion of electric charges. In ordinary material these motions average out to zero, so there is no magnetic field. When some force is applied to the charges to make them have a net motion in a particular direction, we have a net current and a magnetic field. Natural magnetism is due to uncanceled electron spin in the atoms. Again, H has no material dependence—it only depends on the current.

So, D and H can be computed directly from the source charge and current distributions in space. Why do we use E and B? After all, with the right choice of units D and H could be force field vectors. The reason is convenience. When we apply an electric force to a piece of nonconducting material, the atoms in the material respond by separating their charges slightly. Now, in addition to the original source of the applied field, we have the fields of all the spread atomic charges to calculate. Since most materials of practical interest respond in a linear way—that is, the net electric force at a point in the material points in the same direction as the applied electric force and is only reduced in magnitude by the atoms—it is a lot more convenient to say that the net force is equal to the displacement field times a material constant. That's the way it works for magnetic fields, but, again for historical reasons, it's backward for the electric field vectors. The constitutive relations are

$$E = D/\epsilon \tag{2.1}$$

$$B = \mu H \tag{2.2}$$

where ϵ is called *permittivity*, and μ is called *permeability*. In SI units, their values in free space are not 1 but, approximately,

$$\mu_0 = 4\pi 10^{-7}\ \mathrm{H/m}$$

$$\epsilon_0 = \frac{10^{-9}}{36\pi}\ \mathrm{F/m}$$

A given material is usually represented by its relative permeability and relative permittivity. That is, $\mu = \mu_r\mu_o$, $\epsilon = \epsilon_r\epsilon_o$. The relative values are usually equal to or greater than 1. If a material responds to an applied electric field by the atoms' charge-spreading, the fields of the atomic charges oppose the applied field, reducing the net field. Such a material is called a *dielectric*, and you can see in equation (2.1) that a larger ϵ produces a smaller E. So, for such a material, $\epsilon_r > 1$. On the other hand, a ferromagnetic material responds to an applied magnetic field by reinforcing it. This is shown in (2.2) with $\mu_r > 1$. Modern composite materials can show both dielectric and magnetic responses.

Note that in the remainder of the book, I use overbars for general vectors, and the carat (hat) symbol for unit vectors. For example, an electric field vector that points in the θ direction, as shown in the next section, is written as $\bar{E} = \hat{\theta}E_\theta$.

2.1.1 Waves in Space

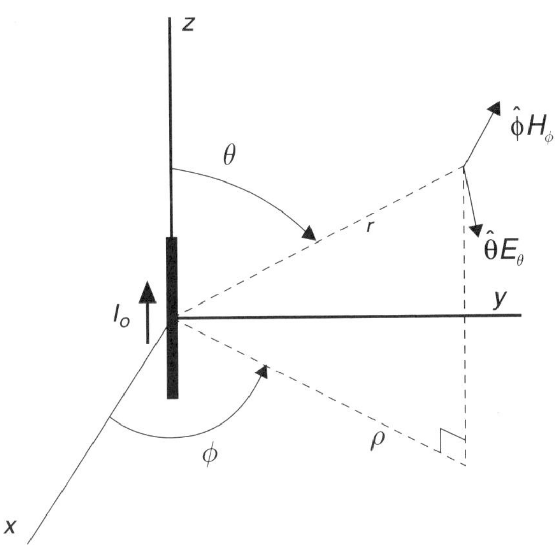

Figure 2.1: A current I_o flows in a wire of length L at the origin of the coordinate system. The field vectors are shown at a point in spherical coordinates, (r,θ,ϕ). The current is sinusoidal in time but independent of position on the wire.

The most elementary real electromagnetic wave is launched by a single-frequency current on an electrically short piece of wire. Such a current is illustrated in Figure 2.1, along with the spherical coordinates needed to describe the wave. Using phasor notation, the wave is usually described by its E and H vectors. Assuming the wire and wave are in an unbounded lossless medium,

$$E_\theta = I_o L \frac{j\omega\mu}{4\pi} \sin(\theta) \frac{e^{-j\beta r}}{r} \tag{2.3}$$

$$H_\phi = E_\theta / \eta \tag{2.4}$$

where

β = radian space frequency = $\omega/v = 2\pi/\lambda$

ω = radian frequency of the current = $2\pi f = 2\pi/T$

v = wave speed in the medium = $1/\sqrt{\mu \in}$

η = wave impedance = $\sqrt{\mu/\epsilon}$

$T = 1/f$ = period of the current in the time domain.

Note the analogy between time domain and space domain quantities.

In the SI system, the units are as follows:

E, newtons/coulomb = V/m

H, A/m

β, radians/m

ω, radians/s

v, m/s

η, Ω

T, s

Some features of (2.3) and (2.4) apply to all radiators, regardless of their size and shape. First, the E and H vectors at any point are perpendicular to each other, and both are perpendicular to the direction from the source to the observation point. Second, the wave phase depends only on r, which means the wave is spherical. Just as when you drop a stone into a still pond, you say the waves are circles because the maxima and minima form circles, so too a snapshot of the wave from a current source would show spherical surfaces for the maxima and minima. Third, the field amplitudes diminish as $1/r$.

Equation (2.3) is grouped into factors that always appear in antenna field expressions. First is the current amplitude. Second is a factor that measures the antenna size in wavelengths. Third is a pattern factor, a factor that depends only on angles. Last is the propagation factor, a function only of distance and which is always the same.

Equations (2.3) and (2.4) represent the wave as having only one vector component for each field. This is due to the choice of the position of the current in the coordinate system. Physically speaking, the electric field vector is always in the same plane as its source current element, and the magnetic field vector is always perpendicular to that plane, always circulating around its source current element. If the orientation of the current element is changed, the mathematical representation of the field vectors must change to reflect these physical facts. If a collection of current elements is being considered, the total field at a point is found first by writing the vector components of the fields due to each current element, and then doing the phasor sum for each vector component. Since we are dealing with electrically small antennas, we discuss only a few simple cases in this text.

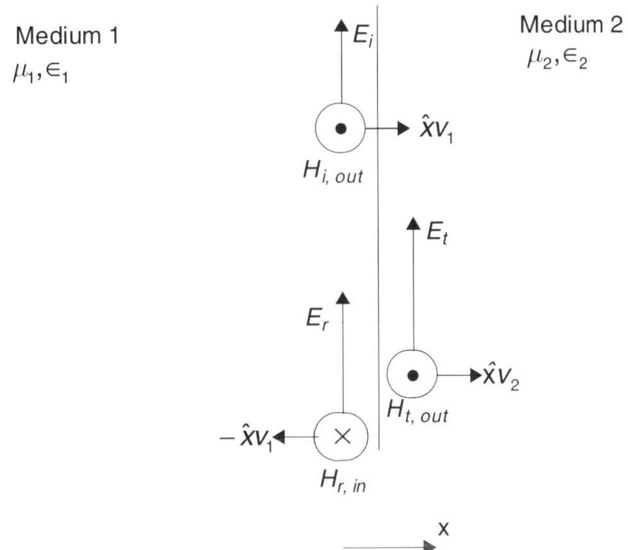

Figure 2.2: Wave reflection at a flat boundary. Medium 1 has the incident and reflected waves, and medium 2 has the transmitted wave. The arrows show the reference directions for the vector fields. Note that total $E = E_i + E_r$ while total $H = H_i - H_r$ in medium 1, all scalar (possibly complex) quantities.

At a distance from the transmitting antenna to the receiving antenna which is large compared to the receiving antenna, the surface of a spherical wave would appear flat because the receiving antenna occupies such a small angular segment. This allows the approximation of radio waves as plane waves for the purpose of what happens at a boundary between two media. The simplest case is illustrated in Figure 2.2, where a plane wave is incident on a plane boundary. In the neighborhood of the boundary, we have an incident wave, a transmitted wave and a reflected wave, all with *E* and *H* vectors parallel to the boundary surface. The reflection occurs because the wave impedances in the two media are not the same, which means there is a mismatch in the ratio of electric to magnetic field values. The total *E* and *H* must each be the same on both sides of the boundary. This leads to

$$E_r = \Gamma E_i = \frac{\eta_2 - \eta_1}{\eta_2 + \eta_1} E_i, \tag{2.5}$$

$$E_t = (1+\Gamma)E_i = \frac{2\eta_2}{\eta_2 + \eta_1} E_i \tag{2.6}$$

Γ is called the *reflection coefficient*. These results are important in their own right, and have a direct analogy with transmission lines. Water is an example of a medium with a low wave impedance compared to air, about 1/9 smaller. For a wave passing from air to water, $\Gamma = -0.8$, which says that the reflected electric vector is reversed in direction from the picture, and is almost as large as the incident field. Only 20% of the incident wave electric field is transmitted, and the total electric field on the air side is reduced to this level. However, the magnetic fields on the air side add, so that the total is larger than the incident.

When a wave passes from a low-impedance to a high-impedance medium, the total electric field on both sides of the boundary is larger than the incident, while the total magnetic fields are smaller. If either medium is electrically conductive or otherwise lossy, the wave impedance and the space frequency become complex. The presence of conductivity will act to reduce the magnitude of the wave impedance. A good electrical conductor, such as a metal, has such a low wave impedance relative to air that it can be treated as a short circuit to the electric field. In this case the boundary surface is an electric field minimum, $E = 0$, and a magnetic field maximum, $H = 2H_i$. These plane wave effects give us a picture to help understand the more complex boundary and body effects discussed and calculated in later sections.

2.1.2 Waves in Transmission Lines

As in the case of waves in space, a time-varying current pushed into a transmission line produces an electromagnetic wave. In this case, however, the wave follows the transmission line, and voltage and current waves are generated on the line. The *E*, *H*, *V*, and *I* waves are as tightly coupled as the *E* and *H* waves in space, so that one can't really say that any one or two cause the rest; they are all caused simultaneously by the applied voltage or current. For a coaxial cable, the waves are entirely enclosed in the dielectric between the outer and inner conductors, and the wave speed is simply determined by

$$v = \frac{c}{\sqrt{\epsilon_r}} \tag{2.7}$$

For twin-lead and microstrip, the fields are partly in the dielectric and partly in air, so the wave speed is between c and the value given by (2.7). The less insulation between the wires in twin-lead, the closer v is to c. With microstrip, ϵ_r in (2.7) is replaced by ϵ_{eff}, for which various formulas are available.

The voltage and current waves are expressed in phasor form as

$$\begin{aligned} V_f(z) &= V^+ e^{-j\beta z}, \quad V_r(z) = V^- e^{+j\beta z} \\ I_f(z) &= I^+ e^{-j\beta z}, \quad I_r(z) = I^- e^{+j\beta z} \end{aligned} \tag{2.8}$$

The subscripts are "f" for forward, "r" for reverse or reflected. The superscripts are "+" for positive z direction and "–" for negative z direction. A schematic transmission line with reference directions and terminations is shown in Figure 2.3. Notice that the positive-going waves have the same negative-going space phase shift as the wave in space traveling out from a current element. This is the phasor version of the $-\beta z$ time domain phase term, the $-z/v$ time delay of the wave going forward. Also note that the origin for the waves is $z = 0$, even though it might be intuitive to have the origin for the reflected wave at $z = d$. Actually, reflections can occur at both ends of the line, so it is better to think of the "r" wave as the reverse wave.

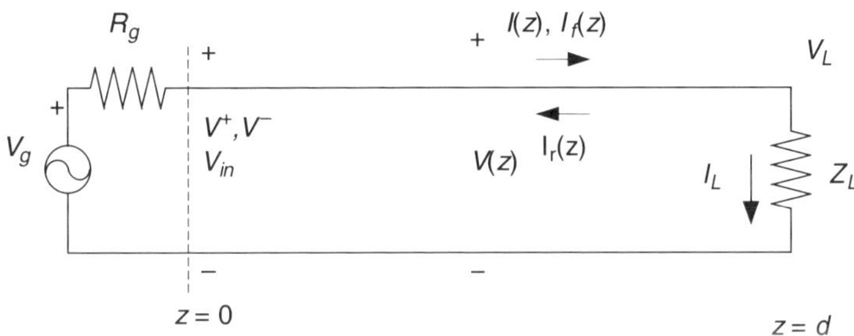

Figure 2.3: Transmission line schematic.

The voltage and current wave amplitudes have a proportional relation just as $\overline{\mathrm{E}}$ and $\overline{\mathrm{H}}$ waves in space. It is usually symbolized by Z_o.

$$V^+ = Z_o I^+,\ V^- = Z_o I^- \tag{2.9}$$

Z_o is usually called the characteristic impedance of the transmission line, but I prefer to call it the wave impedance of the line. At the load, if $Z_L = Z_o$, then $V_L = Z_o I_L$, which is the same current-voltage relation as the forward wave, so no reflection will take place. In this case, the input impedance is just Z_o, so one could define the wave impedance as that value of resistance seen at the input when it is connected at the output of the line.

In general,

$$\begin{aligned} V(z) &= V_f(z) + V_r(z) \\ I(z) &= I_f(z) - I_r(z) \end{aligned} \tag{2.10}$$

At the load, the ratio of total voltage to total current must be Z_L. This boundary condition leads to the definition of a reflection coefficient, and wave relations similar to those for the plane wave and plane boundary case.

$$\Gamma_L = \frac{Z_L - Z_o}{Z_L + Z_o} \tag{2.11}$$

with

$$V^- = V^+ \Gamma_L e^{-2j\beta d} \tag{2.12}$$

and

$$V_L = V^+(1+\Gamma_L)e^{-j\beta d} \tag{2.13}$$

The load voltage plays the same role as the transmitted *E*-field. They each represent the wave or power that is transmitted to the next material object.

The total voltage at any point on the line can be expressed in several ways, depending on the connection one is looking for. Using equations (2.8), (2.10), and (2.12) yields

$$V(z) = V^+(e^{-j\beta z} + \Gamma_L e^{-2j\beta d} e^{j\beta z}) \tag{2.14}$$

The terms can be grouped and interpreted in different ways. The feature I want to point out is that the total voltage is the sum of two phasors of constant magnitude whose phases vary oppositely with z. As one progresses along the line, these phasors will periodically become completely in phase, causing a voltage maximum, and periodically completely out of phase, causing a voltage minimum. The ratio of the voltage maximum to the voltage minimum is called the voltage standing-wave ratio, VSWR, or frequently just SWR. It is related to the magnitude of the reflection coefficient because this magnitude gives the ratio of the magnitudes of the forward and reverse wave voltages.

$$\text{SWR} = \frac{1+|\Gamma_L|}{1-|\Gamma_L|}, \; |\Gamma_L| = \frac{\text{SWR}-1}{\text{SWR}+1} \tag{2.15}$$

Figure 2.4 shows plots of $|V(z)|$ for three cases in which SWR = 2. This corresponds to $|\Gamma| = 1/3$.

With a forward-wave amplitude of 1, the reverse-wave amplitude is 1/3. The maximum voltage is 4/3 and the minimum voltage is 2/3. The curves all have the same shape; their difference is their position on the line. This difference is caused by the phase of the reflection coefficient for the three loads. For $Z_L = 100$, $\Gamma_L = 1/3$, for $Z_L = 40 + j30$, $\Gamma_L = j/3$, and for $Z_L = 25$, $\Gamma_L = -1/3$. Notice that for the two real values of load impedance, $V(d)$ is either the maximum or minimum value, and it

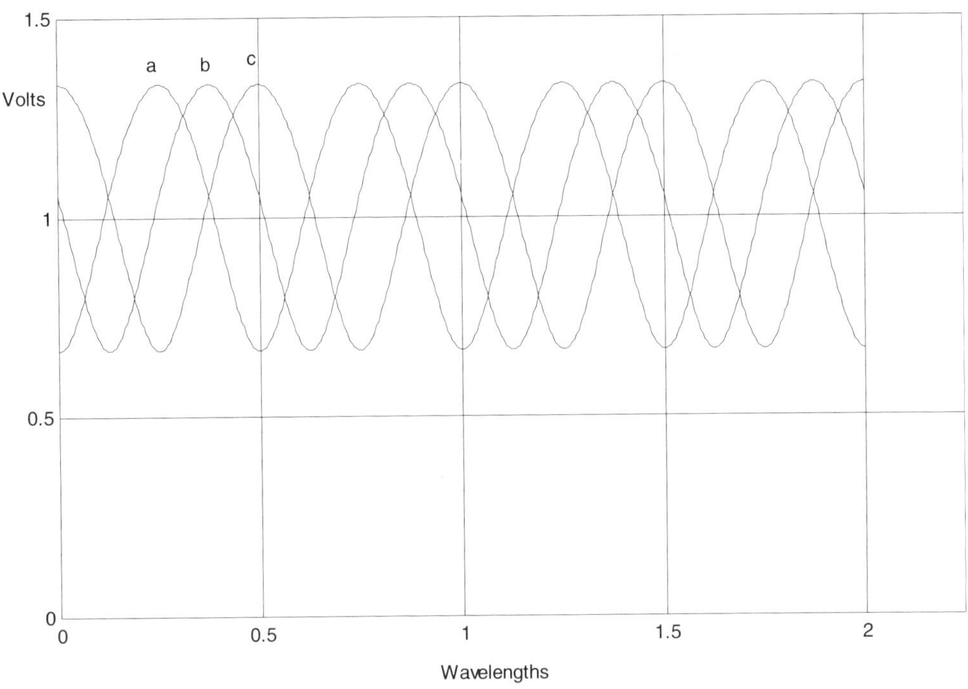

Figure 2.4: Voltage magnitude on two wavelengths of transmission line for VSWR = 2. Z_o = 50Ω. The forward wave amplitude is 1V. (a) Z_L = 25Ω, (b) Z_L = 40 + j30Ω, (c) Z_L = 100Ω.

corresponds to the load value being larger or smaller than Z_o. If the load impedance is a resistance R,

$$\begin{aligned} \text{SWR} &= R/Z_o, \text{ for } R > Z_o \\ \text{SWR} &= Z_o/R, \text{ for } R < Z_o \end{aligned} \tag{2.16}$$

The impedance looking into the line is $V(0)/I(0)$, and the most common expression for it in terms of the line parameters, operating frequency and load impedance is

$$Z_{in} = Z_o \frac{Z_L + jZ_o \tan(\beta d)}{Z_o + jZ_L \tan(\beta d)} \tag{2.17}$$

Sections of transmission line are frequently characterized by their wave impedance and their electrical, rather than physical, length. The *electrical length* is really the wave phase shift, βd, in either degrees or radians. An electrical length of 90° is special. It corresponds to a section of line that is a quarter wavelength. In Figure 2.4, you can see that this is the distance between a voltage minimum and the next voltage maximum on the line. In equation (2.17), the tangents go to infinity, and that leads to

$$Z_{in} = Z_o^2 / Z_L \tag{2.18}$$

Also the two resistance values corresponding to a given SWR are related by this equation, as you might expect.

2.1.3 Power in Waves

From electromagnetic theory, the power density in a wave is

$$S = \frac{|E|^2}{2\eta} = \eta \frac{|H|^2}{2} \ \mathrm{W/m^2} \tag{2.19}$$

assuming peak values for the amplitudes of E and H. In a lossless medium the magnitude of either field decreases as $1/r$, so the power density decreases as $1/r^2$. A source with no angle variation in its wave is called *isotropic*, and is a convenient fiction for theory. At a distance r from an isotropic source, the surface area of a sphere is $4\pi r^2$, so multiplying the constant power density by the spherical area gives a total power independent of distance. This is an illustration of what you would expect in a lossless medium; at a given distance from a source the power density integrated over the corresponding spherical surface gives the total power, a constant.

Returning to the plane waves at a flat material boundary, the power densities associated with the incident and reflected waves are

$$S_i = \frac{|E_i|^2}{2\eta_1}, \quad S_r = \frac{|E_r|^2}{2\eta_1} = |\Gamma|^2 \frac{|E_i|^2}{2\eta_1} = |\Gamma|^2 S_i \tag{2.20}$$

and the transmitted power density must be

$$S_t = (1 - |\Gamma|^2) S_i \tag{2.21}$$

Similar-looking relations apply at the transmission-line load. The power in each wave is

$$P_f = \frac{|V^+|^2}{2Z_o}, \quad P_r = \frac{|V^-|^2}{2Z_o} = \frac{|\Gamma_L V^+|^2}{2Z_o} = |\Gamma_L|^2 P_f$$
$$P_L = \left(1 - |\Gamma_L|^2\right) P_f \tag{2.22}$$

Here we are talking about total power, not power density. Since we are assuming a lossless transmission line, the power in the waves is the same all along the line. Also the net power into the line must equal the load power. The following example illustrates some of the ideas presented so far.

Example 2.1 A Quarter-Wave Matching System

Suppose we have a transmitter rated at 100W into 50Ω that we want to use to drive an antenna whose input impedance is 12.5Ω. We plan to do this with a quarter-wave section of transmission line. From (2.18), $Z_o = \sqrt{50 \times 12.5} = 25\Omega$. Since the load is resistive, we know from (2.16) that the SWR is 2, and from (2.11) that $\Gamma_L = -1/3$. For a load power of 100W, the load voltage must be $V_L = \sqrt{2 \times 12.5 \times 100} = 50\text{V}$ peak or 35 Vrms. From (2.22), the forward wave power must be $100/(1 - (1/3)^2) = 112.5$W, and the reflected power is $112.5 \times (1/3)^2 = 12.5$W. This checks with $P_f - P_r = P_L$. The corresponding voltage amplitudes are $V^+ = \sqrt{2 \times 25 \times 112.5} = 75\text{V}$ and $V^- = V^+/3 = 25$V. At the load, these amplitudes subtract to give the needed 50, and at the transmitter end, they add to give 100V. Is this correct? The input power is $P_{in} = V_{in}^2/(2R_{in}) = 10^4/(2 \times 50) = 100$W, as required.

We can trace this through in a slightly different and more general way. Since $\beta d = \frac{2\pi}{\lambda} \cdot \frac{\lambda}{4} = \frac{\pi}{2}$, the forward-wave voltage at the load is $V_f(d) = V^+ e^{-j\pi/2} = -jV^+$. Likewise, the reverse-wave voltage is $V_r(d) = jV^-$. Since the reflection coefficient is $-1/3$, $jV^- = (-jV^+)(-1/3)$ or $V^- = V^+/3$. We can use the 100W requirement either at the load or at the input to get the total voltage as above, and solve for the wave amplitudes. At the input, we know the total voltage has to be $100 = V^+ + V^- = 4V^+/3$ or $V^+ = 75$ and $V^- = 25$. Then at the load $V_L = V_f(d) + V_r(d) = (-j75) + (j25) = -j50$V. I took phase reference at the input. If I had used the output power requirement and assumed the output voltage to be real, the wave amplitudes would have had a j in front of them. Taking phase reference at the input will always mean that the input voltage and wave amplitudes will be real numbers. Notice the effects of the waves' phase shifts along the line. By choice, at the input the wave amplitudes are real numbers, going toward the load the forward wave loses phase and the reverse wave gains phase. In this case, on a quarter-wave-length section, the forward wave loses 90° (π/2) while the reverse wave gains 90°. In a more general case, we could only assume one of the wave amplitudes to be a real number because a complex reflection coefficient or a different line length will cause a phase shift between the wave amplitudes, as can be seen in (2.12), but the forward wave will lose phase and the reverse wave gain phase in the same manner.

2.2 Polarization

The orientation of a wave in space is an important matter. It affects the ability of a receive antenna to convert wave power to circuit power. To begin, suppose you are an observer standing on the *x-y* plane ($z = 0$) and you use a receiver with a whip antenna to detect a wave from a vertical transmitting antenna. Assume that there is no radiation from any other source on your frequency, including passive re-radiation. Since the total wave is made up of waves from current elements that all point in the *z* direction, the total wave will have an electric field vector that points in the *z* direction in your neighborhood. You will observe a maximum signal when your whip is vertical. The total $\overline{E}$ vector always points in the same direction, so the wave is called *linearly polarized*. Because the orientation of the $\overline{E}$ vector in your coordinate system is vertical, it is also called *vertically polarized.*

Suppose now you can travel around and up and down above the *x-y* plane. At any point, you will find the maximum signal when your whip antenna is in a plane containing the transmit antenna. Since at a given point your whip orientation doesn't have to change in time to get the maximum signal, and if you turn away from the maximum-signal position you can find a position that gives no signal, the wave is still linearly polarized. Because the $\overline{E}$ vector is still in a plane that is perpendicular to the ground, the wave is still said to be vertically polarized.

If now the transmitting antenna is a straight horizontal wire, the wave it launches will have a horizontal $\overline{E}$ vector. Again, if you move your whip around while you are standing on the *x-y* plane, you will find the maximum signal when the whip is horizontal and zero signal when the whip is vertical. So the wave is still linearly polarized because the direction of the $\overline{E}$ vector doesn't change in time at a particular location. And now it is called *horizontally polarized* because the $\overline{E}$ vector is everywhere horizontal.

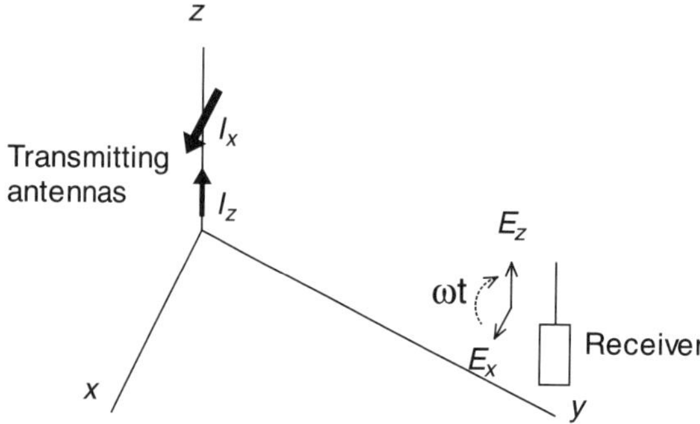

Figure 2.5: Illustration for elliptical polarization. The receiver is assumed to be at least several wavelengths from the transmitting antennas.

Now suppose the transmitter drives two antennas, one a horizontal wire and the other a vertical wire, as illustrated in Figure 2.5. Suppose the horizontal wire is parallel to the *x* axis and above it some distance. If the currents are in phase, and you are standing on the *y* axis, the total wave in your neighborhood can be represented as $\overline{\text{E}}_{\text{total}} = \hat{x}E_x + \hat{z}E_z$. This is a vector with constant amplitude pointing in a specific direction that is neither horizontal nor vertical. So now we can only call it linearly polarized, because turning your whip in a plane parallel to the *x-z* plane will still find a position of maximum signal and a position of no signal.

Next, suppose the transmitting system puts a phase shift of ϕ radians into the horizontal wire's current. This phase shift will appear in the wave radiated by the horizontal wire. Now the total $\overline{\text{E}}$ vector in your neighborhood on the *y* axis is

$$\overline{\text{E}}_{\text{total}} = \hat{x}E_x e^{j\phi} + \hat{z}E_z \qquad (2.23)$$

The two components are no longer in phase so direct vector addition isn't possible. What does this expression mean? We need to go back from the phasor domain to the time domain.

$$\overline{\text{E}}_{\text{total}} = \hat{x}E_x \cos(\omega t + \phi) + \hat{z}E_z \cos(\omega t) \qquad (2.24)$$

The first thing to observe is that there is no time in the radio-frequency cycle that the total $\overline{\text{E}}$ is zero. This means that no matter where you put your whip antenna in a plane parallel to the *x-z* plane, it will always have a signal to detect. Going through one RF cycle, there are two times when the $\overline{\text{E}}$ vector is purely *x*-directed and two other times when it is purely *z*-directed. In between, the vector direction progresses around the *y* axis. This general situation is called *elliptical polarization*. In Figure 2.5, the dashed-line curved arrow shows the direction of rotation of the $\overline{\text{E}}$ vector for $\phi > 0$. From the transmitter's point of view, the direction is counterclockwise, or left-handed. A wave in which the $\overline{\text{E}}$ vector rotates this way is called *left-handed elliptically polarized* (LHEP). If $E_x = E_z$ and $\phi = \pi/2$, the ellipse becomes a circle and the wave is called LHCP. Figures 2.6 and 2.7 show some examples of the path traced by the total $\overline{\text{E}}$ vector at a point in space, through one time cycle.

Elliptical or circular polarization can be deliberate, as in this example, or accidental. Many satellite signals of interest are transmitted as circularly polarized waves, so receiving antennas to capture the most power from these waves should be circularly polarized also.

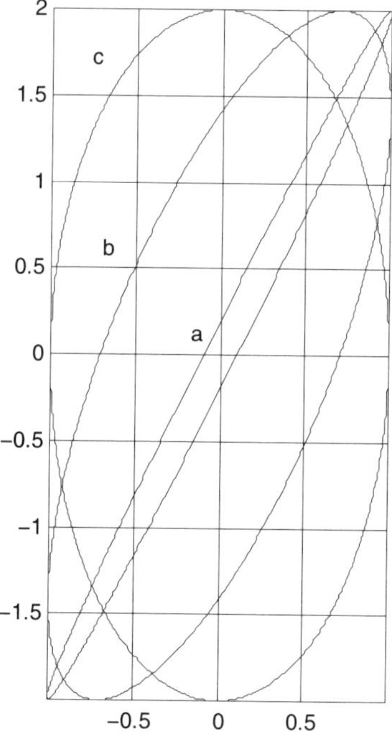

Figure 2.6: Plots of the z component vs. the x component through one RF cycle for equation (2.24). E_x = 1, E_z = 2 for all curves. (a) ϕ = 0.1 (b) ϕ = $\pi/4$ (c) ϕ = $\pi/2$.

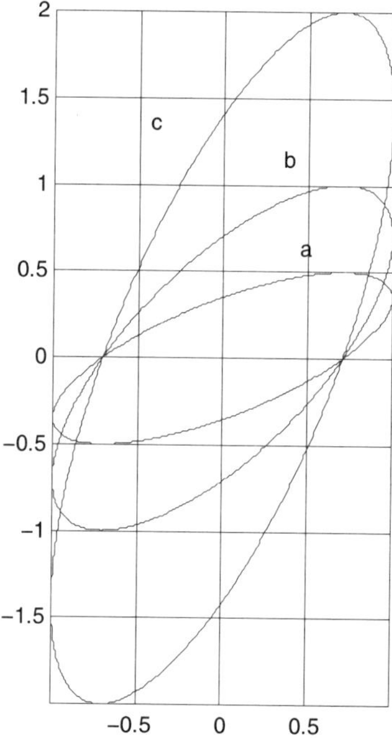

Figure 2.7: Plots of the z component vs. the x component for one RF cycle for (2.24). ϕ = $\pi/4$, E_x = 1. (a) E_z = 0.5 (b) E_z = 1 (c) E_z = 2.

2.3 The Short Dipole

To begin, consider a two-wire transmission line as illustrated in Figure 2.8(a). The line is driven by a signal source at its left and is open-circuited at its right end. The open-circuit causes a standing-wave pattern on the line, with zero current at the open end. At a given cross-section of the line, the currents are equal in magnitude and opposite in direction. Their radiated fields will likewise be equal in magnitude and opposite in direction at any point in space, providing the wire separation is tiny compared to wavelength. Therefore, the line produces no net radiation. Now suppose the ends of the wires are turned up and down, respectively, as shown in Figure 2.8(b). The radiated fields from current on these turned-out wires no longer cancel because the currents now have the same space orientation. The current still has to be zero at the open ends of the dipole. If $L << \lambda$ the current distribution will be the tail end of a sinusoid as shown in Figure 2.8(c). The current distribution will not be exactly as it was on the transmission line in 2.8(a) because the current in each leg of the dipole has to work against its own field to generate the radiated power, but the sinusoid approximation is useful. An expression for the current amplitude that can be derived from the transmission line equations with $\Gamma_L = 1$ is

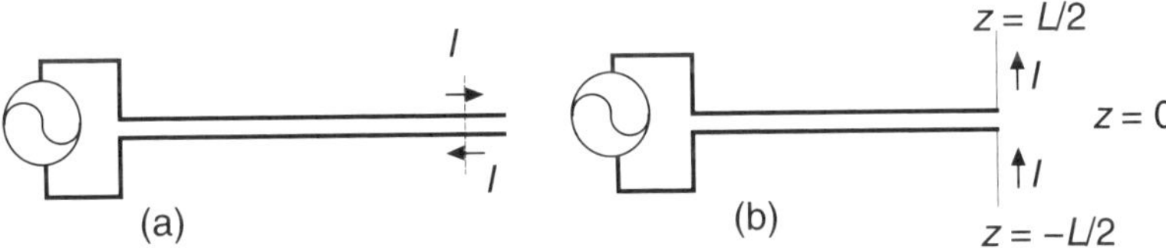

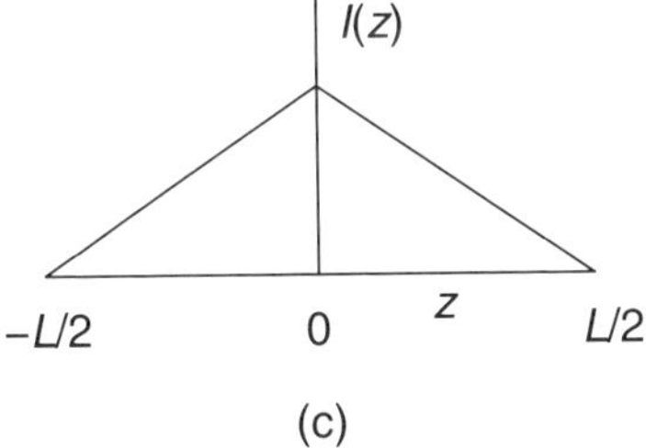

Figure 2.8: (a) A two-wire line with an open end. (b) The line with the ends turned out to form a dipole. (c) Current on the dipole.

$$I(z) = I_o \sin\left(\frac{2\pi}{\lambda}\left(\frac{L}{2} - |z|\right)\right) \Big/ \sin\left(\frac{\pi L}{\lambda}\right) \tag{2.25}$$

where I_o is the current at the dipole-transmission line junction. Remember that this is the amplitude at a point z of a time-varying single-frequency current. Equation (2.3) gives the radiated field for a uniform distribution of current. This may be adapted to the present situation by replacing L with dz and I_o with $I(z)$ to give the field due to a differential length of wire. The total field is the sum of the individual fields, and this sum goes to an integral as dz gets very small.

$$E_\theta = \frac{j\eta}{2\lambda}\sin(\theta)\int_{-L/2}^{L/2} I(z)\frac{e^{-j\beta R}}{R}dz \tag{2.26}$$

$R = |\hat{r}r - \hat{z}z|$ is the distance from dz to the field point. If $L << r$, and $L << \lambda$, then R can be treated as a constant in both the exponential phase term and the $1/R$ term. Essentially, R can be replaced by r. Making these items constants lets us take them outside the integral, leaving only $I(z)$. Multiplying and dividing by L gives an expression for the average current, and the field expression reduces to

$$E_\theta = \frac{j\eta L}{2\lambda}\sin(\theta)\frac{e^{-j\beta r}}{r}I_{average} = \frac{j\eta L}{2\lambda}\sin(\theta)\frac{e^{-j\beta r}}{r}\alpha I_o \tag{2.27}$$

It will be convenient for later use to symbolize the ratio of average to terminal current as α. For the $I(z)$ of (2.25),

$$\alpha = \frac{I_{average}}{I_o} = \frac{1 - \cos\left(\frac{\pi L}{\lambda}\right)}{\frac{\pi L}{\lambda}\sin\left(\frac{\pi L}{\lambda}\right)} \tag{2.28}$$

In the limiting case, this becomes just

$$\alpha = 1/2 \tag{2.29}$$

2.3.1 Radiation Pattern

Radiation pattern is the name given to a plot of the angle function in the field expression of an antenna. There are a number of ways this information can be presented, and you will run across most of them in one place or another. In general, we can call this function $f(\theta,\phi)$ where (θ,ϕ) are the spherical coordinate angles in Figure 2.1. Typically, we are interested in the magnitude of this function, and it is usually normalized so that its maximum value is 1. Define a pattern function for the amplitude as

$$f_E = |f(\theta,\phi)|/f_{\max} \tag{2.30}$$

Power density is proportional to amplitude squared, and a *power pattern* function is sometimes used. The power pattern function is just the square of the amplitude pattern function:

$$f_P = (f(\theta,\phi)/f_{\max})^2 \tag{2.31}$$

If the pattern is expressed in decibels (dB) there is no difference between amplitude and power patterns.

$$f_{\text{dB}} = 20\log_{10}\left(|f(\theta,\phi)|/f_{\max}\right) = 20\log_{10}\left(f_E\right) = 10\log_{10}\left(f_P\right) \tag{2.32}$$

Since this is the most common form, we use it in the following examples.

Once the data form is decided, the next issue is the presentation of the data. The following two plots show the sin(θ) angle function for our vertical short dipole in the commonly used rectangular and polar formats. A problem with using dB is that small values of f_E produce large negative values of f_{dB}. To get around this, the data range is limited by some minimum value, and all values smaller than the limit are replaced by the limit. This is sometimes called clipping. The plots in Figures 2.9 and 2.10 are clipped at –30 dB. A problem with the polar format is the reader's perception. There is a tendency to regard the polar display as a distance-covered plot, which it isn't. What the plot is telling you in both formats is that if you travel in a vertical circle with the dipole at the center, your relative readings of amplitude or power in the radiated field will fit these patterns. In this particular case, the dipole field doesn't vary with the horizontal angle, ϕ, so the result is the same for any vertical circle. Furthermore, if we walk around the dipole in horizontal circles, the pattern is just a circle in polar format or a horizontal line in rectangular format, an uninteresting result.

In Figure 2.10, the angle layout is generic so the z axis is horizontal, and a full circle is displayed even though the range of θ is just 0 to 180°. While θ and ϕ are the standard angles in math and science and a lot of antenna theory, for practice on or near the earth's surface, azimuth and elevation angles are frequently used. *Azimuth* is the same as ϕ, and *elevation* is 90° – θ.

Things get more interesting if we consider a horizontal dipole. For a dipole lying on the x axis the field expression is, from equations (2.27) and (A.12),

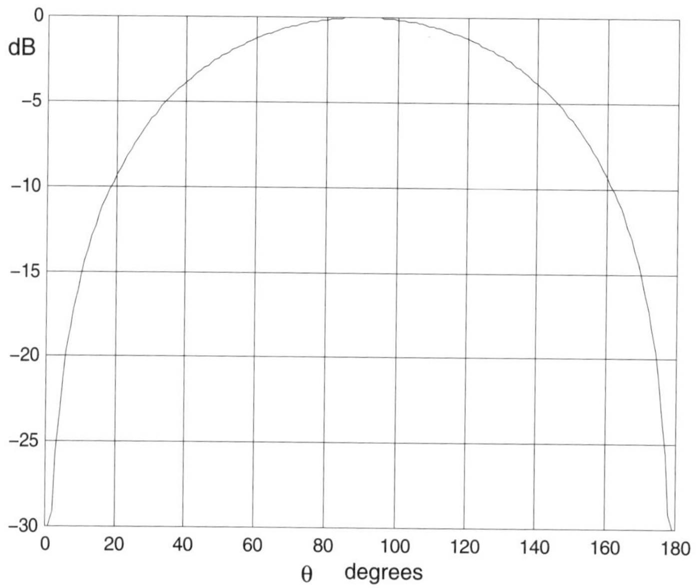

Figure 2.9: Pattern plot for vertical dipole in rectangular coordinates.

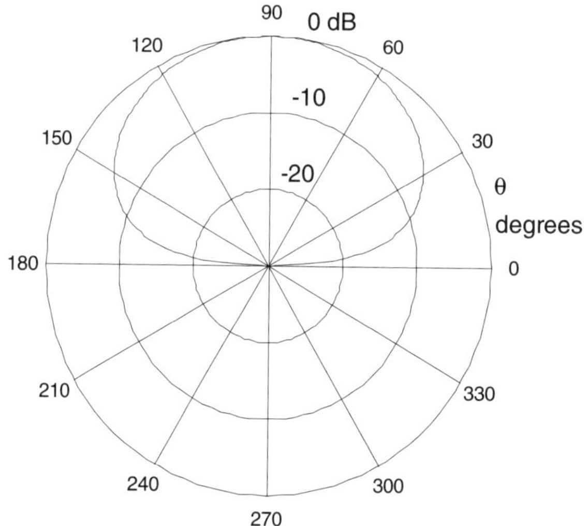

Figure 2.10: Polar plot version of Figure 2.9. The plot center corresponds to –30 dB.

$$\bar{E} = \frac{j\eta L}{2\lambda}\frac{1}{2}\alpha I_o\left[\hat{\phi}\sin(\phi) - \hat{\theta}\cos(\phi)\cos(\theta)\right]\frac{e^{-j\beta r}}{r} \quad (2.33)$$

The two vector components are in phase, so the polarization is still linear, as it should be, and the amplitude is the square root of the sum of the squares of the vector components. It is more convenient to work with the power, avoiding the square root.

$$f_P = \sin^2(\phi) + \cos^2(\phi)\cos^2(\theta) \quad (2.34)$$

When the dipole was on the z axis, the pattern in any vertical plane was half a figure eight in polar format. Now with the dipole on the x axis, we should expect the same pattern in any plane containing the x axis. However, it is customary to measure and plot patterns of azimuth for constant elevation, and elevation for constant azimuth, meaning that 3D space is sliced in planes parallel to the x-y plane for azimuth plots, and in planes containing the z axis for elevation plots. Figure 2.11 shows azimuth plots for the short dipole on the x axis, with elevation as a parameter. You can see that the only curve with nulls is the 0°-elevation plot. This is the only one containing the x-y plane. Likewise, in Figure 2.12 the only elevation curve with a real null is the 0°-azimuth plot because it is the only one containing the x axis. Fundamentally, the only places you can't get a signal from a dipole is when you're looking directly end-on.

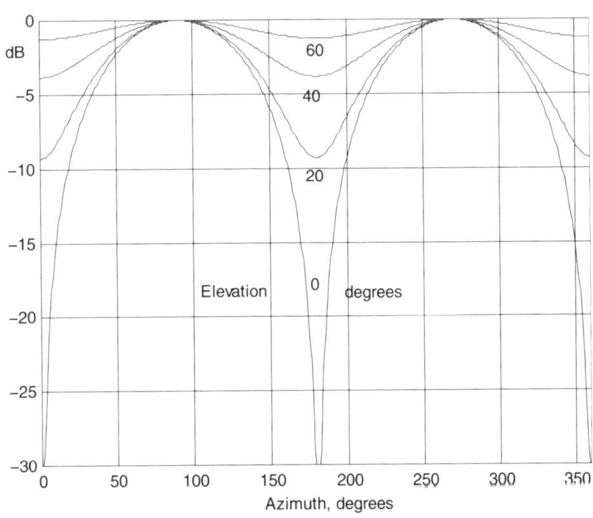

Figure 2.11: Azimuth pattern plots for the *x*-axis short dipole.

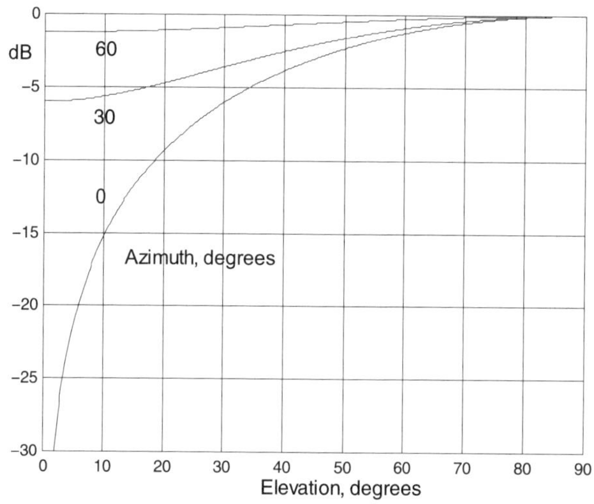

Figure 2.12: Elevation patterns for a small dipole on the *x* axis.

2.3.2 Circuit Behavior

In general, at a single frequency, an antenna can be modeled as a voltage source, a pair of series resistors, and a series reactance as shown in Figure 2.13. The voltage source represents the conversion of a passing wave to terminal voltage, discussed later in this chapter. One resistor, R_{rad}, is the circuit model that represents the power radiated away in a transmitted wave, the other resistor, R_{loss}, models power loss to heat in the antenna conductors. The reactance, jX, models the energy stored in the antenna's near field, just as in an ordinary inductor or capacitor. This model applies whether the antenna is being used to transmit or receive.

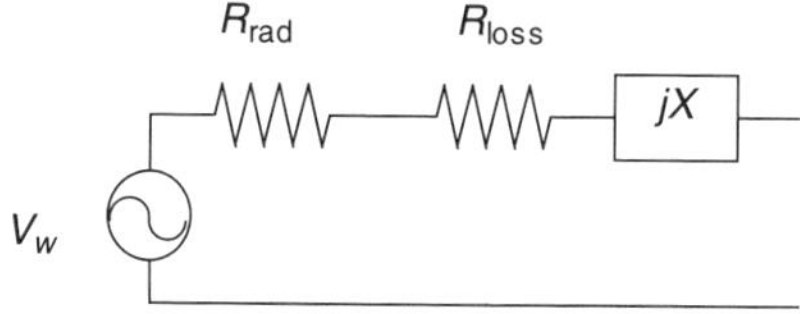

Figure 2.13: Equivalent circuit for an antenna.

As described in section 2.1.3, the radiated power is the total power in the wave at a constant distance from the source. This is the integral of the power density over a sphere's surface. The element of surface area in spherical coordinates is

$$dA = r^2 \sin(\theta) d\theta d\phi \quad \textbf{(2.35)}$$

The radiated power for any antenna can be expressed as

$$P_{rad} = I_o^2 M \int_0^{2\pi} \int_0^{\pi} f_P(\theta,\phi) \sin(\theta) d\theta d\phi \quad \textbf{(2.36)}$$

in which I_o is the peak terminal current, f_P is the power pattern function described in the last section, and M holds all the other factors in the power density expression other than $1/r^2$, which is cancelled in dA. For the short dipole, approximately

$$M = \frac{\eta}{8}\left(\frac{\alpha L}{\lambda}\right)^2 \tag{2.37}$$

The power pattern integral is 8π/3. This leads to

$$P_{\text{rad}} = \frac{2\eta\pi}{3}\left(\frac{\alpha L}{\lambda}\right)^2 \frac{I_o^2}{2} = 80\left(\frac{\alpha\pi L}{\lambda}\right)^2 \frac{I_o^2}{2} \tag{2.38}$$

in free space where $\eta = 120\pi$. So the radiation resistance is

$$\begin{aligned} R_{\text{rad}} &= 80\left(\frac{\alpha\pi L}{\lambda}\right)^2 \\ &= 20\left(\frac{\pi L}{\lambda}\right)^2, \quad \textit{for an open-ended short dipole.} \\ &= 80\left(\frac{\pi L}{\lambda}\right)^2, \quad \textit{for a dipole with uniform current.} \end{aligned} \tag{2.39}$$

You can see that, unlike an ordinary resistor, the radiation resistance is a strong function of wavelength. The two special cases for $\alpha = 0.5$ and 1 are extremes that can be approximated in practice.

At frequencies above the audio band, the current in a wire conductor is not uniform. Its amplitude decreases exponentially from the outside. This behavior is called skin effect or current crowding. An equivalent surface thickness is usually defined as the *skin depth*,

$$d_5 = \sqrt{\frac{2}{\omega\mu\sigma}} = \sqrt{\frac{1}{\pi f \mu\sigma}} = \sqrt{\frac{\lambda}{\pi c\mu\sigma}} \tag{2.40}$$

If the smallest cross-sectional dimension of the wire is at least six skin depths, the per-unit-length resistance may be approximated as

$$R_{\text{pu}} = \frac{1}{\sigma d_p d_s} = \frac{1}{d_p}\sqrt{\frac{\omega\mu}{2\sigma}} = \frac{1}{d_p}\sqrt{\frac{\pi\eta}{\lambda\sigma}} = \frac{1}{d_p}\sqrt{\frac{\pi f\mu}{\sigma}} = \frac{1}{d_p}R_{sq} \tag{2.41}$$

where d_p is the distance around the perimeter, and R_{sq} is called the *resistance per square* or surface resistance. Notice that neither the radiation resistance nor the loss resistance scale with wavelength, unlike the radiated field amplitude.

Since the current on the dipole is a function of position, we can't just multiply R_{pu} by the dipole length to get R_{loss}. Instead, we have to do a power loss calculation as we did for radiated power. The power lost in a small length dz is $I^2(z)R_{pu}dz/2$, so the total loss is

$$P_{loss} = \frac{1}{2}R_{pu}\int_{-L/2}^{L/2} I^2(z)dz \approx R_{pu}\int_0^{L/2}\left(I_o\left(1-\frac{2z}{L}\right)\right)^2 dz = R_{pu}\frac{L}{3}\cdot\frac{I_o^2}{2} \tag{2.42}$$

from which

$$R_{loss} = R_{pu}\frac{L}{3} \tag{2.43}$$

I have modeled the dipole current as a triangle function, which is a reasonable approximation for $L < \lambda/5$.

The calculation of X is much more complicated. For one thing, it depends on the detailed geometry of the source-antenna junction, not just the antenna geometry itself. A result that ignores the geometry and assumes the triangle current distribution is

$$X = -\frac{\eta\lambda}{\pi^2 L}\left(\ln\left(\frac{L}{d}\right)-1\right) \tag{2.44}$$

where d is the wire diameter. This expression is adapted from references [1] and [2]. From this expression, we can find the dipole capacitance as

$$C = \frac{\epsilon\pi L}{2}\Bigg/\left(\ln\left(\frac{L}{d}\right)-1\right) \tag{2.45}$$

Example 2.2 **Dipole Input Impedance and Efficiency**

Suppose we wish to use a dipole for the 30-MHz citizens' band. We choose its length as 1 m and its diameter as 12 mm, and it's made of aluminum tubing. The wavelength is 10 m, so $L/\lambda = 0.1$. From (2.39), $R_{rad} = 1.9739\Omega$. The skin depth, from (2.40) and assuming $\sigma = 26$ MS/m for aluminum, is $d_s = 18.02$ μm. The wall thickness for any reasonable tubing should be greater than 60 μm so the skin-effect model should be okay. The per-unit-length resistance from (2.41) is $R_{pu} = 0.056614\Omega$, and $R_{loss} = R_{pu}/3 = 0.01887\Omega$. Since the loss resistance is only 1% of the radiation resistance, you would expect this short dipole to be very efficient, and it is, by itself.

The reactance from (2.44) is $X = -(1200/\pi)(\ln(1/0.012)-1) = -1{,}307\Omega$. This value corresponds to a capacitance of 4.06 pF. The ratio $|X|/(R_{rad} + R_{loss}) = 656$. A good tuning coil might have an $X/R_{\text{coil-loss}} = 200$. This would imply $R_{\text{coil-loss}} = 6.54\Omega$. As you can see, such a series resistance would absorb 3.3 times more power making heat than the antenna would radiate.

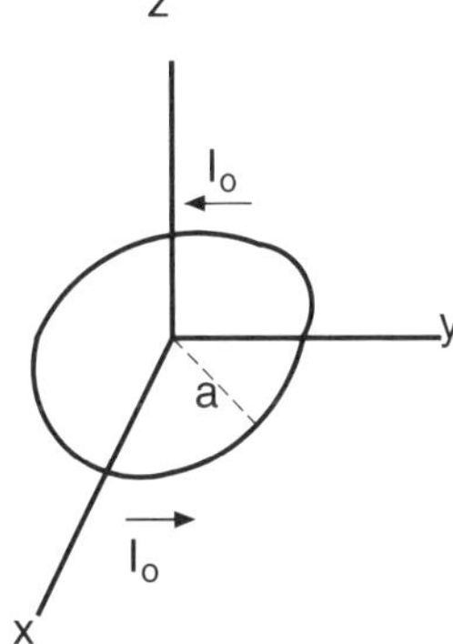

Figure 2.14: A circular loop with uniform current lying in the *x*-*y* plane.

2.4 The Small Loop

A circular loop with terminal current I_o, lying in the x-y plane, is shown in Figure 2.14. The loop is considered small [3] if its circumference is less than $\lambda/3$, which implies a diameter less than about $\lambda/10$. In this case, it is a good approximation to say that the current is the same all the way around the loop. That means that at the ends of any diameter there are equal currents going in opposite directions. This makes a situation between the transmission line and the dipole, in that the fields generated by these currents tend to cancel, but not completely. Starting from (2.3) and using results from Appendices A and B, and using symmetry and smallness, I show in Appendix C that the radiated field is

$$\overline{\mathrm{E}} = \eta\pi\frac{A}{\lambda^2}I_o\sin(\theta)\hat{\phi}\frac{e^{-j\beta r}}{r} \tag{2.46}$$

in which A is the loop area. This result is valid for a small loop of any shape. The horizontal small loop has the same amplitude pattern as the vertical dipole, but is horizontally polarized instead of vertically polarized. Some authors call them dual antennas, and some authors call the loop antenna a magnetic dipole. It is true that in the near field the loop stores most of its energy in a magnetic field and the short dipole stores its near-field energy in an electric field, but the waves radiated by each have the same E/H; they are equally electric and magnetic. The near-field energy storage makes their circuit behavior different, but as far as their antenna behaviors are concerned, there is nothing particularly magnetic or electric about them.

The field strength varies inversely with λ^2 instead of λ. This is due to two effects—the basic $1/\lambda$ scaling for length, and another $1/\lambda$ scaling for separation between the opposing current elements.

2.4.1 Circuit Behavior

The first resonance of a circular loop of diameter D is at about $\pi D/\lambda = 0.49$ [4], slightly lower for thicker loops, slightly higher for thinner loops. This has an effect even at frequencies where the loop is considered to be electrically small, so the effect can be modeled by adding a shunt capacitor to the loop's circuit model. This is shown in Figure 2.15.

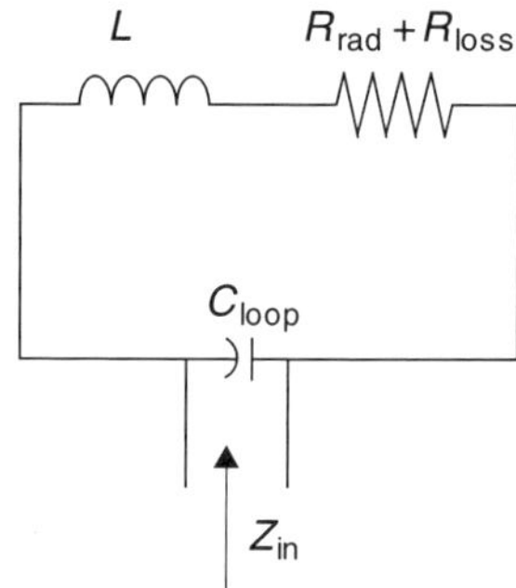

Figure 2.15: Equivalent circuit for a small loop.

To get at the value of C_{loop}, it is sufficient to use the usual relation for the resonant frequency of a simple RLC circuit, $C_{\text{loop}} = 1/(\omega^2 L)$. For the present case,

$$0.49 = \frac{\pi D}{\lambda} = \frac{\pi f D}{c} = \frac{\omega D}{2c}, \quad \omega = \frac{0.98c}{D}$$

This leads to

$$C_{\text{loop}} = \frac{D^2}{(0.98\text{c})^2 L} = 11.57\frac{D^2}{L} \quad \text{pF} \tag{2.47}$$

when D is in meters and L is in μH.

The radiated power density for the horizontal small loop is

$$\text{S} = \frac{|\overline{\text{E}}|^2}{2\eta} = \eta\left(\pi\frac{A}{\lambda^2}\right)^2 \frac{I_o^2}{2}\sin^2(\theta)\frac{1}{r^2} \tag{2.48}$$

The power pattern integral is the same as for the dipole, 8π/3, so the radiation resistance is

$$R_{\text{rad}} = \eta \frac{8\pi^3}{3}\left(\frac{A}{\lambda^2}\right)^2 \tag{2.49}$$

In free space or air, where $\eta = 120\pi$,

$$\begin{aligned} R_{\text{rad}} &= 320\pi^4\left(\frac{A}{\lambda^2}\right)^2 = 20\left(\beta^2 A\right)^2 \\ &= 0.346 f^4 A^2 \end{aligned} \tag{2.50}$$

In the last expression f is in MHz and A is in m^2.

The resistance per unit length is the same expression as for the dipole, but the total loss resistance is just the loop length time the resistance per unit length because the loop current doesn't vary with position.

$$\begin{aligned} R_{\text{loss}} &= R_{\text{pu}} l = \frac{l}{\sigma d_p d_s} = \frac{l}{d_p}\sqrt{\frac{\pi f \mu}{\sigma}} \\ &= \frac{l}{d_p} 1.987 \sqrt{\frac{f}{\sigma}} \quad m\Omega \end{aligned} \tag{2.51}$$

The last line is for f in MHz and σ in MS/m. d_p is the perimeter of the wire or tubing and l is its length.

The small loop is an inductor and there are many expressions available for its inductance. Following are some adapted from reference [5, pp. 52–53]. All shapes are made of round wire of diameter d, and the equations are the high-frequency versions.

Circular loop of diameter D:

$$L = \frac{\mu}{2} D\left[\ln\left(\frac{8D}{d}\right) - 2\right] \tag{2.52}$$

Square loop of side s:

$$L = \frac{2\mu}{\pi} s\left[\ln\left(\frac{2s}{d}\right) + \frac{d}{2s} - 0.774\right] \tag{2.53}$$

Rectangular loop, sides s_1, s_2, diagonal g, and wire length W.

$$L = \frac{\mu}{\pi} W \left\{ \frac{s_1}{W} \ln\left(\frac{4 s_1 s_2}{d(s_1 + g)} \right) + \frac{s_2}{W} \ln\left(\frac{4 s_1 s_2}{d(s_2 + g)} \right) + \frac{2g - d}{W} - 1 \right\} \tag{2.54}$$

Simplified expression for some regular shapes of wire length W:

$$L = \frac{\mu}{\pi} W \left[\ln\left(\frac{4W}{d} \right) - K \right] \tag{2.55}$$

$K = 2.451$ for a circle,

$K = 2.853$ for a square,

$K = 3.197$ for an equilateral triangle,

$K = 3.332$ for an isosceles right triangle.

I've written (2.54) in a more complex form than necessary in order to make it more easily comparable to the other forms. Any length unit can be used, as long as μ is converted correctly. The value of K in (2.55) decreases as the number of sides in the loop increases because the area enclosed by a fixed length of wire increases.

Example 2.3 **Loop Impedance and Efficiency**

Let's return to the conditions of Example 2.2, and try a circular loop of the same 1-m diameter and 12-mm aluminum tubing as the dipole of that example. The loop area is $\pi/4$ m^2. Using (2.50), the radiation resistance is

$$R_{\text{rad}} = 320\pi^4 \left(\frac{\pi/4}{100} \right)^2 = \frac{20\pi^6}{10^4} = 1.92278\Omega$$

The wire length is π, and $R_{\text{pu}} = 0.056614\Omega$ as with the dipole, so $R_{\text{loss}} = 0.1778\Omega$. This is almost 10% of the radiation resistance, so we can expect a bit over 90% efficiency for the antenna itself.

For inductance calculations it is often convenient to use $\mu = 0.4\pi$ µH/m. From (2.55)

$$L = \frac{0.4\pi}{2\pi} \pi \left(\ln\left(\frac{4\pi}{0.012} \right) - 2.451 \right) = 0.2\pi(6.954 - 2.451) = 2.829\ \mu\text{H}$$

The inductive reactance is $X = 2\pi 30L = 533\Omega$.

From (2.47), $C_{loop} = 11.57/2.829 = 4.09$ pF. This makes the net input impedance for Figure 2.15 $Z_{in} = 6.06 + j905.5\Omega$. The loop would be tuned with a capacitor that is nearly 100% efficient. So far, the loop looks like a better system choice.

2.5 Directionality, Efficiency, and Gain

It was mentioned earlier that an isotropic source is one that radiates the same field strength in all directions. Our small antennas don't quite do that, as they have nulls in their radiation patterns. One of the purposes of building electrically large antennas and arrays (collections of antennas tied to the same source) is to concentrate as much of the radiated power as possible in one direction. Therefore it is of interest to have a measure of the directionality of an antenna and there are several in common use.

Some of these relate to features of the antenna pattern, and are not of much relevance to our simple antennas. However, some low-profile antennas exhibit more interesting patterns, so I will define some of these terms here.

Front-to-Back Ratio

For this term to make sense, there has to be a direction that is the "front" or direction of intended transmission. The *front-to-back ratio* (F/B) is the ratio of the power density in the intended direction to that radiated to the "back," 180° around from the front. For the loop or the dipole, $F/B = 1$, or 0 dB.

Half-Power Beamwidth

The *half-power beamwidth* (HPBW) requires both an intended direction and a plane of definition. For example, suppose we are at the center of coordinates and we want to transmit along the y axis with a horizontal dipole. We would orient the dipole along the x axis to get the most power density in the y direction. Now if we consider the azimuth pattern at zero elevation, we have a maximum in the $+y$ direction and in the $-y$ direction, and nulls on the x axis. Since the $+y$ direction is intended, the part of the pattern between the $+y$ axis and the nulls on the x axis is called the *main lobe* or beam. At two azimuth angles in this pattern, the power density falls to ½ the maximum. HPBW is the difference between these angles in the azimuth plane. HPBW can also be defined in the elevation plane, and for our example it has no meaning because there is no change in power density in the y-z plane.

Side Lobe Level

If the pattern in a particular plane has more than a front lobe and a back lobe, the other lobes are called *side lobes*. The maximum side lobe level (SLL), relative to the main lobe, is sometimes of interest.

Aside from these feature-based parameters, there is a more basic measure of directionality called *directivity*. It is defined as

$$D = \frac{\text{Maximum power density}}{\text{Average power density}} = \frac{4\pi}{\int_0^{2\pi}\int_0^{\pi} f_p(\theta,\phi)\sin(\theta)d\theta d\phi} \tag{2.56}$$

where f_p is the power pattern function defined in section 2.3.2. We have already seen this integration when we calculated the total radiated power in order to find the radiation resistance. For the small loop and dipole, the integration is 8π/3 so $D = 3/2$ or 1.76 dB, in free space. When the antenna is not in the "free space" of our imagination, other factors come into play and one has to be careful to find out exactly what is meant by a claimed directivity or gain. This is discussed in section 2.9, *Ground Effects*.

Using the radiation resistance and the directivity, we can write an expression for the maximum radiated power density:

$$S_{max} = \frac{I_o^2}{2} R_{rad} \frac{D}{4\pi r^2} \tag{2.57}$$

where I_o is the peak terminal current.

As shown in the previous two examples, antennas radiate power, and they also lose power to heat. For the series circuit model, efficiency, the ratio of the radiated power to the total input power, is just the ratio of the radiation resistance to the total resistance.

$$\text{Efficiency} = Eff = \frac{R_{rad}}{R_{rad} + R_{loss}} \tag{2.58}$$

This is straightforward for the short dipole, but I have complicated the issue for the loop by adding a shunt C to account for the first resonance. The effect of this is examined in Chapters 3 and 9.

Antenna *gain* is the product of efficiency and directivity:

$$G = D\ Eff \tag{2.59}$$

Starting from input power, the maximum radiated power density is:

$$S_{max} = \frac{P_{in} Eff}{4\pi r^2} D = \frac{P_{in} G}{4\pi r^2} \tag{2.60}$$

Antenna gain can be thought of as the actual maximum power density over the ideal (lossless) average power density. While the two small antennas in the previous two examples have equal directivity, the loop has less gain because it has lower efficiency.

References

[1] Kazimierz Siwiak, *Radiowave Propagation and Antennas for Personal Communications*, Artech House, 1995, p. 233.

[2] Wen Geyi, "A Method for Evaluation of Small Antenna Q," IEEE Trans. on Antennas and Propagation, vol. 51, no. 8, p. 2127, August, 2003.

[3] John D. Kraus, *Antennas*, 2nd ed., p. 2 55, McGraw-Hill, 1988.

[4] Richard C. Johnson, Editor, *Antenna Engineering Handbook*, 3rd ed., p. 5–11, McGraw-Hill, 1993.

[5] Frederick Terman, *Radio Engineers' Handbook*, McGraw-Hill, 1943.

Chapter 2 Problems

Section 2.1.1

2.1 Write a time-domain expression equivalent to (2.3).

2.2 Given a current of 1A flowing on a wire of length 1 m having a frequency of 1 MHz, what is the magnitude of the $\overline{\text{E}}$ and $\overline{\text{H}}$ fields at 10 km?

2.3 Suppose the current element of problem 2.2 lies on the x axis. At an observation point 10 km out on the y axis, what are the x,y,z components of the $\overline{\text{E}}$ and $\overline{\text{H}}$ fields?

2.4 For the current element of problems 2.2 and 2.3, suppose the observation point is at $r = 10$ km, $\phi = \pi/3$ and in the x,y plane. Find the ϕ,θ components of $\overline{\text{E}}$ and $\overline{\text{H}}$.

2.5 Derive expressions similar to (2.5) and (2.6) for the $\overline{\text{H}}$ fields.

2.6 Assume that for water $\epsilon_r = 81$. For a wave in air with $E_i = 1$ mV/m, normally incident on a flat water surface, find all the other $\overline{\text{E}}$ and $\overline{\text{H}}$ wave fields, and the total $\overline{\text{E}}$ and $\overline{\text{H}}$ fields at the boundary.

2.7 For a good conductor, the wave impedance magnitude is $|\eta| = \sqrt{\frac{\omega\mu}{\sigma}}$, where σ is the conductivity. Suppose a normally incident wave with $E_i = 1$ mV/m and frequency 100 MHz is blocked by a layer of copper with conductivity 57 MS/m. Find the total $\overline{\text{E}}$ and $\overline{\text{H}}$ fields at the boundary.

Section 2.1.2

2.8 Find an expression for the total current $I(z)$ similar to (2.14).

2.9 If $|V_f| = 20$ mV and $\Gamma_L = 3$, find the other voltage and current wave amplitudes at the load end of a 75Ω line. Find the load voltage and current. What is Z_L?

2.10 Derive (2.17) by writing equations for $V(0)$ and $I(0)$.

2.11 Derive an expression similar to (2.17) except using Γ_L instead of Z_L.

2.12 Find Z_o and βd for a transmission line to match $Z_L = 40 + j30$ to $Z_{in} = 75\Omega$.

2.13 Find Z_o and βd given Z_L and R_{in}.

2.14 A section of line is only a quarter wavelength at a single frequency. Suppose a line is needed to match 5Ω to 45Ω at 1 GHz. Find its wave impedance, Z_q. Find the range of frequencies for which SWR<2. You may do this by writing a program to compute SWR vs. frequency, or you may solve the problem analytically.

2.15 The bandwidth of a matching system can be broadened by using two quarter-wave lines to change the impedance level by smaller ratios. Suppose then that we want a system that goes from R_{lo} through a line with Z_{q1} to R_{mid}, then through a line with Z_{q2} to R_{hi}. If $R_{mid}/R_{lo} = R_{hi}/R_{mid}$, find the wave impedances in terms of the termination resistances.

2.16 Write a program to find and plot VSWR vs. normalized frequency for a two-line matching system as described in problem 2.15. Given $R_{lo} = 12.5\Omega$, $R_{hi} = 200\Omega$, find the normalized bandwidth for VSWR<1.5.

Section 2.1.3

2.17 Derive (2.21) starting from wave amplitude expressions.

2.18 Suppose we have a transmission line with $Z_o = 50\Omega$, $d = 3/8\ \lambda$ at the operating frequency, terminated in $Z_L = 40 - j30\Omega$. If $P_f = 100$W, find P_r, P_L, and the waves, input and load voltages.

Section 2.2

2.19 Find the value of ωt that produces the maximum $|E|$ for each curve in Figure 2.6, and the corresponding $|E|$.

2.20 Repeat Problem 2.19 for Figure 2.7.

2.21 Find a general expression for the angle ωt that produces the maximum $|E|$ for an elliptically polarized wave.

Section 2.3

2.22 Derive (2.25), the sinusoidal current distribution on a short dipole.

2.23 Write a program and plot (2.25) for $L = 0.05\lambda$, 0.1λ, and $L = 0.25\lambda$.

2.24 From equation (2.28), write a program and plot $\alpha = I_{average}/I_o$ vs. L/λ for $0 < L/\lambda < 0.5$.

Section 2.3.1

2.25 (a) Derive the power pattern function for a short dipole on the y axis.
(b)Write a program and generate azimuth plots.

Section 2.3.2

2.26 Show that the power pattern integral in (2.36) is $8\pi/3$ for the short dipole.

2.27 Justify the approximation and the final result in (2.42).

2.28 Equation (2.44) probably isn't useful for $L/d < 10$. Suppose we increase the tubing diameter in Example 2.2 to 0.1 m. Recalculate X, C, $Q = |X|/R$, and the system efficiency with a tuning coil whose Q is 200.

Section 2.4.1

2.29 Rework the relevant parts of Example 2.3 with the tube diameter increased to 0.1 m.

2.30 Rework Example 2.3 with the loop diameter reduced to 0.5 m. The tubing diameter is still 12 mm.

Section 2.5

2.31 Justify equation (2.60) for maximum power density.

2.32 As mentioned in the section on polarization, antennas can be driven by a single source with phase shifters inserted to achieve a specific purpose. For this problem, consider two vertical dipoles located on the y axis at $+d$ and $-d$. Their purpose is to achieve maximum signal in the $+y$ direction. This is accomplished by inserting electrical phase shifters so that the two waves are in phase in this direction. See Figure P2.32. The wave from I_1 is advanced in space phase by $\beta d\sin(\phi)$ and the wave from I_2 is delayed in space phase by $-\beta d\sin(\phi)$. To get the two waves in phase on the y axis, opposite electrical phases of βd are inserted into the signals supplying the currents. The electric field in the x-y plane is modified by the factor $2\cos(\beta d(\sin(\phi)-1)$. For $d = 1/8\ \lambda$, $\frac{1}{4}\ \lambda$, and $3/8\ \lambda$, plot the power patterns in the x-y plane, and find F/B and HPBW.

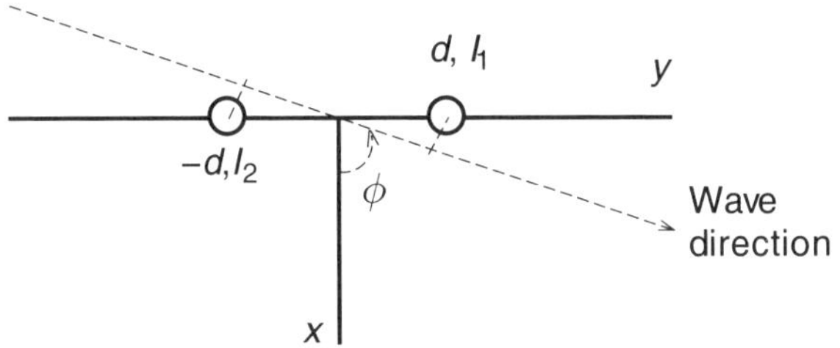

Figure P2.32: Two vertical dipoles on the *y* axis.

2.33 Consider again the antennas of Figure P2.32. It can be shown from Appendix B that the pattern function for this arrangement is:

$$f_p = \sin^2(\theta)\cos^2\left(\beta d\left(\sin(\theta)\sin(\phi)-1\right)\right)$$

Use numerical integration to find the directivities for $d = 1/8\lambda$, $\frac{1}{4}\lambda$, and $3/8\lambda$.

CHAPTER 3

Antenna Fundamentals II

3.1 Bandwidth and Quality Factor, Q

In the early days of electronic radio, the ability of a receiver to select just one station out of the many with sufficient strength to be detected was determined by the quality of a coil. Figure 3.1 shows a simplified schematic of an RF amplifier. To begin, we assume that the following stage has a very high input impedance so that the load on the active device (a vacuum tube in those days) is just the elements shown. Here r_c is the resistance inherent in the coil wire. An ideal (perfect) coil would have no loss resistance. The voltage out of the stage is:

$$V_{\text{out}} = -g_m Z(j\omega) V_{\text{in}} \tag{3.1}$$

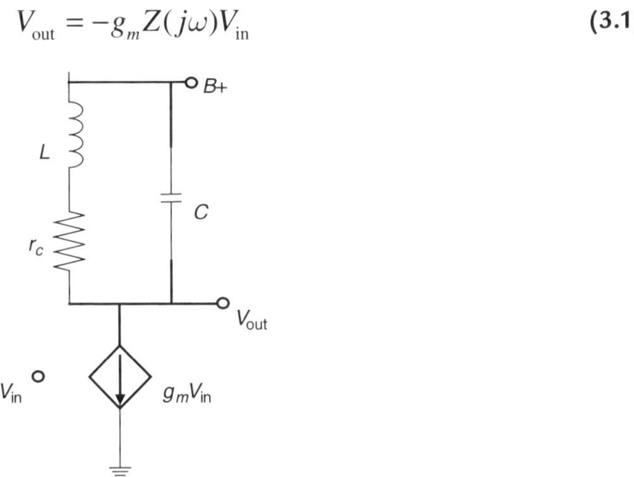

Figure 3.1: Simplified schematic of a radio receiver amplifier stage.

We may further assume that the power into a load further into the receiver is proportional to the square of the output voltage of our stage. This output power is:

$$P = G g_m^2 \left|Z(j\omega)\right|^2 \left|V_{\text{in}}\right|^2 \tag{3.2}$$

where G is a constant representing the power gain of the following stages. Let P_m be the maximum power as a function of frequency, and Z_m be the maximum magnitude of $Z(j\omega)$. Then the normalized power is:

$$P_{\text{norm}} = \frac{P}{P_m} = \frac{|Z(j\omega)|^2}{Z_m^2} \tag{3.3}$$

This normalized power is similar in concept to a gain known as *transducer power gain (TPG).*

$$\text{TPG} = \frac{\text{Load power}}{\text{Maximum power available from the source}} \tag{3.4}$$

When we talk about antenna bandwidth and bandwidth of an antenna-matching network combination, it is with respect to this gain.

Ordinarily we think that the maximum impedance of a parallel-resonant circuit is at the frequency where the impedance is entirely real and the imaginary part goes to zero. If the resistance were in parallel with the coil instead of in series, this would be true, but in this case it is not. There are two frequencies of interest then, the frequency at which we have maximum impedance magnitude, and the resonant frequency. In any well-designed stage, these are very close, but by carefully studying these issues now we will have some results we can use later when discussing impedance-matching.

The impedance function is

$$Z(j\omega) = \frac{r_c + j\omega L}{1 - \omega^2 LC + j\omega C r_c} \tag{3.5}$$

By straightforward calculus and algebra, you can find that the frequency for $|Z| = Z_m$ is the basic resonant frequency expected for the series circuit,

$$\omega_0^2 = \frac{1}{LC} \tag{3.6}$$

Define $Q_0 = \dfrac{\omega_0 L}{r_c}$. This is the *quality factor* of the coil at frequency ω_0. Then

$$Z_m^2 = \left(Q_0^2 + 1\right) Q_0^2 r_c^2 \tag{3.7}$$

The normalized half-power bandwidth is:

$$\frac{\omega_H - \omega_L}{\omega_0} \approx \frac{1}{Q_0} \tag{3.8}$$

The frequency at which $Z = R_r$ is real can be found by equating the ratios of imaginary to real parts for the numerator and denominator of Z.

$$\omega_r^2 = \frac{1}{LC} - \frac{r_c^2}{L^2} = \omega_0^2\left(1 - \frac{1}{Q_0^2}\right) \tag{3.9}$$

$$R_r = \frac{L}{r_c C} = Q_0^2 r_c = \left(Q_r^2 + 1\right) r_c, \quad Q_r = \frac{\omega_r L}{r_c} \tag{3.10}$$

We see from these results that the quality factor of the coil determines both the maximum output of the stage and its effect on the bandwidth. The coil's Q is also the *unloaded Q* of the circuit.

Now suppose that the following stage has a finite input resistance, R_i. Effectively, this resistor is in parallel with C, and the two together could be treated as a lossy capacitor. For a parallel model like this, the quality factor is the susceptance over the conductance.

$$Q_{\text{shunt}} = \frac{|B|}{G} = \frac{\omega C}{G_i} = \omega C R_i \tag{3.11}$$

The effective Q of the circuit is now reduced, and is approximately

$$Q_e = \frac{Q_{\text{shunt}} Q_0}{Q_{\text{shunt}} + Q_0} \tag{3.12}$$

This is also called the *loaded Q* of the resonant circuit. Loss increases bandwidth, $\text{BW} = 1/Q_e$.

Example 3.1 **Effects of Coil Q and Loading**

For this example I chose easy numbers. The coil resistor is 1Ω, $L = 20$ μH, $C = 0.05$ μF, and $\omega_0 = 1$ Mrad/s. This makes $Q_0 = 20$ and the value of R_r is 400Ω. I wrote a MATLAB function to take these values and a radian frequency vector and a load resistor, and compute the power vector, max impedance magnitude, and unnormalized half-power bandwidth. The frequency interval used was 1 krad/s. The normalized power curves are shown in Figure 3.2. Notice that the half-power frequency on the high side of ω_0 is further out than the one on the low side. This is caused by using a linear frequency scale even though circuit behavior is dependent on frequency ratio.

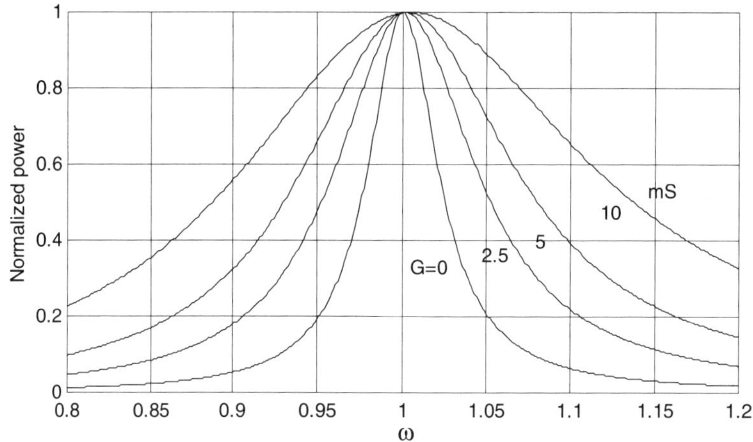

Figure 3.2: Power plots for Example 3.1. *G* is the applied load conductance.

The following values were computed by the program. If the arithmetic average of the half-power frequencies in each case were used to normalize the bandwidth, the results would have been slightly smaller than those in the table, because of the upward shifts of the frequency responses.

Table 3.1: Computed maximum impedance and bandwidth for various loads, Example 3.1 and Figure 3.2.

Load R, Ω	Z_{max}, Ω	Bandwidth, Mrad/s	Q_{shunt}	Q_e	$1/Q_e$
None	400.5	0.049	---	20	0.05
400	200.25	0.099	20	10	0.1
200	133.5	0.149	10	20/3	0.15
100	80.1	0.249	5	4	0.25

The RLC circuit we've been talking about so far is a particular kind of electromagnetic resonator. It turned out that a measure of the coil quality also predicts the circuit bandwidth. In the 1940s other kinds of resonators, shorted transmission line sections and waveguide cavities, came along. In all these resonators, energy is stored in, and oscillates between, electric fields and magnetic fields. At resonance, the power source supplies the resonator losses and the total stored energy is constant. These devices have no easy lumped-element description, so a more general definition of Q that covers all cases to that point was introduced. From [1, p. 269, eq. (152)],

$$Q = 2\pi \frac{\text{Energy stored}}{\text{Power lost per cycle}} \tag{3.13}$$

The discussion in [1] above this equation implies that the equation was developed in order to predict the bandwidth in the same manner as with the lumped-element resonators. Since "Power lost per cycle" is just the average power times the period and $T = 1/f$, this definition of Q is more commonly written

$$Q = \omega \frac{\text{Energy stored}}{P_{\text{average}}} \tag{3.14}$$

Again, this is unloaded Q.

There must have been a discussion of Q for small antennas in the 1940s, because L. J. Chu published a paper [2] in 1948 that found an expression for the minimum achievable Q for a small antenna. The difference between resonators and antennas is that a good resonator has a high Q, whereas a good antenna has a low Q. That is, unless you're trying to use the antenna to filter transmitter harmonics. Chu expressed the fields of a dipole with arbitrary current distribution in terms of spherical waves and developed an equivalent circuit for each mode of the wave. For each mode, he gives a Q expression that is twice (3.14), because the antenna is not a resonator by itself, but must be tuned with an external element that has an equal amount of stored energy of the opposite type [2, p. 1169, eq. (12)]. On p. 1170, he applies his (12) to the entire antenna structure, and states his assumptions that the antenna is internally resonant and conductively lossless.

$$Q = 2\omega \frac{\text{Mean electric energy stored}}{\text{Power radiated}} \tag{3.15}$$

Chu then states that the minimum Q for a small antenna occurs when only the TM_1 or TE_1 or both modes are excited, but does not give an explicit expression for this value. However, other investigators have done so, for instance [3, p. 171]. The Chu Q limit for a linearly polarized antenna that fits inside a sphere of radius r is:

$$Q_{\text{Chu}} = \frac{1 + 3\beta^2 r^2}{\beta^3 r^3 \left(1 + \beta^2 r^2\right)} \rightarrow \frac{1}{\beta^3 r^3}, \; r \ll \lambda \tag{3.16}$$

The limiting value (lower bound) on Q has been reexamined several times over the years, and the most recent expression as of this writing is [4, p. 2118] not much different.

$$Q = \frac{1}{\beta^3 r^3} + \frac{1}{\beta r} \tag{3.17}$$

The Q can also be expressed in terms of antenna input impedance. Chu gave an expression for the antenna below series resonance, and other people have generalized this expression to work through series resonance [5, eq. (17)].

$$Q = \frac{1}{2R_{\text{in}}}\left(\omega\frac{dX_{\text{in}}}{d\omega} + \left|X_{\text{in}}\right|\right) \tag{3.18}$$

Remember that we are still talking about unloaded Q and relating it to the half-power bandwidth.

Q is important as a figure of merit for comparing different antennas, and half-power bandwidth has application importance. There are three problems with the expressions given so far. The first is that the input resistance is not constant with frequency. The second is that some small antennas work near parallel resonance (also called antiresonance). The third is that transmitters generally can't stand to run at full power at the SWR that corresponds to the half-power frequency. These issues have been addressed in [6].

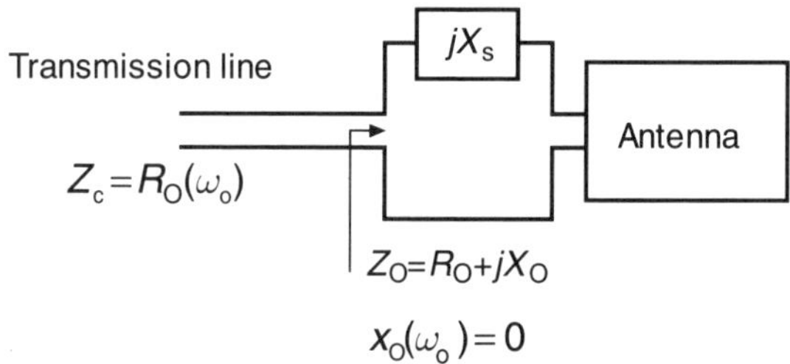

Figure 3.3: Set-up for matched fractional bandwidth definition.

A system in which the antenna is series-resonated by an external reactor at the desired center frequency ω_o is hypothesized and illustrated in Figure 3.3. The transmission line is present in order to define a reflection coefficient at its load end, the tuned system input. Using the fact that a specified SWR also specifies the reflection coefficient magnitude, and assuming $Q > 4$ and certain other size assumptions, the authors derive an expression for the *matched fractional bandwidth* between the points where the specified SWR is met.

$$BW_{\text{SWR}} = \frac{\Delta\omega_{\text{SWR}}}{\omega_o} = \frac{2bR_O(\omega_o)}{\omega_o\left|\dfrac{dZ_O(\omega_o)}{d\omega}\right|}, \quad b = \frac{\text{SWR}-1}{\sqrt{\text{SWR}}} \tag{3.19}$$

If we call the antenna input impedance Z_A and bear in mind that $R_O(\omega_o) = R_A(\omega_o)$,

$$\frac{dZ_O(\omega_o)}{d\omega} = \frac{dR_A(\omega_o)}{d\omega} + j\frac{dX_s(\omega_o)}{d\omega} + j\frac{dX_A(\omega_o)}{d\omega} \tag{3.20}$$

This expression also gives the power bandwidth if the source happens to have an impedance equal to the wave impedance of the hypothetical transmission line.

In [6], the authors define Q as in equation (3.14), but including the series reactance in the energy calculation. The result is:

$$Q(\omega_o) = \frac{b}{BW_{\text{SWR}}} = \frac{\omega_o \left| \dfrac{dZ_O(\omega_o)}{d\omega} \right|}{2R_O(\omega_o)} \qquad \textbf{(3.21)}$$

The authors assert that (3.19–21) work in all frequency ranges as long as $Q > 4$. Because (3.21) is based on (3.14), it gives the unloaded Q and $1/Q$ gives the unloaded half-power bandwidth of the antenna and reactor as a resonator. For a matched source and transmission line, load SWR = 5.828 for half power.

Example 3.2 **SWR Bandwidth of a Lumped-Element Resonator**

Consider again the lumped resonator of Example 3.1, only this time think of it as an equivalent circuit for an antenna. We again have a 1-Ω resistor in series with 20 μH, the pair shunted with 0.05 μF. Assume initially that we want SWR = 5.828. It is possible to find analytical expressions for the derivatives in (3.19–20), but I will take you through the numerical process as if the impedance data were only available as numbers. To approximate the derivative numerically, we must sample frequency with small steps either side of the nominal operating frequency. In the present case, $\omega_o = 1$, so frequency steps of 0.01 seems reasonable. First, I must generate the antenna impedance data at 0.99, 1, and 1.01 Mrad/s. The results are

ω	Z_A
0.99	$351.34 + j121.04$
1	$400 - j20$
1.01	$338.49 - j154.53$

At ω_o, $R_A = 400$, $X_A = -20$. Therefore the series resonator is an inductor, and its reactance is $X_s = \omega 20/\omega_o$. To calculate the approximate derivatives, I use the top and bottom entries of the data, so they are centered at ω_o: $d\omega = 0.02$, $dR_A = -12.844$, $dX_A = -275.57$, $dX_s = 0.4$, $dZ_O = dR_A + j(dX_A + dX_s) = -12.844 - j275.14$. Note that even though the internal resistor is a constant, its effect at the terminals is not. The differential is about 5% of that for the system reactance. $|dZ_O| = 275.47$. $|dZ_O|/d\omega = 13{,}774$. From (3.19), $BW = 800b/13{,}774$. For SWR = 5.828, $b = 2$, so $BW = 0.116$. This result agrees closely with the previous example when the resonator was shunted with a 400-Ω load, which would be the matched source impedance in the present case.

Suppose now we want to limit the SWR to 2, a common specification. For $\text{SWR} = 2,\ b = 1/\sqrt{2},\ \text{BW} = 0.041$. From 12% to 4% bandwidth.

Example 3.3 Parallel-Tuned Loop SWR Bandwidth

Consider next the loop antenna of Example 2.3. It has a much higher reactance-to-resistance ratio than the last example, but the resistance is a strong function of frequency. How do you expect the derivatives will play out?

To begin, I found the Q of the antenna as an inductor, not considering C_{loop}. At 30 MHz, $\omega_o = 60\pi = 188.5$ Mrad/s, $Q = \omega_o L/(R_{\text{rad}} + R_{\text{loss}}) = 533.25/2.10058 = 253.86$, so the matched half-power bandwidth is about $2/Q = 0.0078783$. Next I'll parallel-tune the antenna. The admittance of the loop's series branch at 30 MHz is $Y = 7.387–j1875$ μS. The shunt C must be 1875/188.5 = 9.9485 pF. The added capacitance is $C_{\text{tune}} = 9.9485–C_{\text{loop}} = 5.859$ pF. The tuned input resistance is $1/\text{real}\{Y\} = 10^6/7.387 = 135.37$ kΩ. Since I know the loop resistance and reactance values at ω_o, I can find them at other frequencies by scaling them according to their frequency variation behaviors.

$$R_{\text{rad}} = R_{\text{rad}}\left(\omega_o\right)\left(\frac{\omega}{\omega_o}\right)^4, \quad R_{\text{loss}} = R_{\text{loss}}\left(\omega_o\right)\sqrt{\frac{\omega}{\omega_o}}, \quad X = \frac{\omega}{\omega_o}X\left(\omega_o\right) \tag{3.22}$$

I don't really need the X scaling since I also have the inductance, but it's there for completeness. I found I had to try successively smaller frequency steps to get a stable result. Writing it as a MATLAB vector, $\omega = \omega_o$*[1 – d,1,1 + d]. I got the same bandwidth to four figures for $d = 1e - 5$ and $d = 1e - 6$. For $d = 1e - 6$, $dZ_O = –0.3868 – j137.47$, $|dZ_O| = 137.47$ so the resistance change made no real difference. The half-power matched bandwidth came out to 0.007878. As a check, I ran the normalized transmitted power calculation from 29.7 to 30.3 MHz using 10-Hz steps. The half-power bandwidth came out to 0.007878.

In this case, the main conclusion is that differentiating the reactance of the antenna and tuning element captures the effect of stored energy just as well as using the reactance of one of the system elements. The effect of the frequency variation of the loop resistance still isn't strong enough to change the result.

3.2 Impedance Matching and System Efficiency

The terminal impedance of small antennas isn't usually what is needed for the rest of the system. In fact, most electronic subsystems, in their basic forms, don't have the right input and output impedances to work together directly. Therefore, almost all subsystems are either designed to have I/O impedances that meet a system standard, or they must be interfaced with separate impedance-matching circuits. Impedance-matching is a wide and deep subject in the literature, and is closely related to filters, both in theory and practice. The area can be divided conveniently four ways, narrow-band vs. wide-band designs, lumped-element (LC) vs. distributed-element (transmission-line section) designs. Reference [7] has an extensive presentation of LC narrow-band methods. Reference [8] is a classic handbook for all four categories. [9] contains a modern survey emphasizing transmission-line solutions. References [10–12,15] are articles that move ideas between LC and short-line solutions. Reference [13] is a classic text emphasizing broadband LC theory, and [14] is a presentation emphasizing broadband transmission-line network theory. In the following subsections, I present only a beginning to the subject, enough to give you some appreciation of what's involved, and some useful methods and examples.

In general, the purpose of an impedance-matching network is either to interface a source to a load so as to allow maximum power transfer, or to present a load to a source with a value that allows the source to develop maximum power. These are not the same thing, or equivalent. For example, the basic low-frequency output impedance of a BJT is very high, and it can be modeled as a current source. If you match its output impedance with a high-impedance load, full allowed voltage swing will only get you a tiny current swing, so no significant power is developed. What the BJT needs to develop maximum power is a load that allows rated current at rated voltage, or something close. On the other hand, in order to get maximum signal power from an antenna into a receiver, the Maximum Power Transfer Theorem applies, so the antenna impedance must be converted to the nominal system impedance, which is usually 50Ω in the U.S. In practice, the power amplifier load problem is handled by the amplifier design engineer(s) who must convert the system-level impedance to a value suitable for the amplifier output stage. In all applications, the job of the antenna engineer(s) is to convert the antenna's impedance to the system-level value, regardless of whether the antenna is loading a transmitter or driving a receiver. The approach of using a system-level impedance that everything has to work with is good for interchangeability, but frequently produces more parts than would be needed for a fully integrated design.

3.2.1 Narrow-Band Matching

"Narrow-band" really means "single-frequency," a design that's done for one frequency that works well enough over a small band of frequencies around the design frequency. The match is perfect at the center frequency, and degrades away from it. At a single frequency, the matching network must accomplish two things: it must eliminate the load reactance and shift the resistance level to the system value. To accomplish two things, you need two variables. The quarter-wave line in Example 2.1 only had to do one thing—change the resistance level—and it had only one variable, its wave impedance. This points to the idea that if you made both the line length and wave impedance variables, you could do impedance-matching with just a section of transmission line. This is true, but is seldom done because it doesn't necessarily produce the most compact design. At frequencies below about 1 GHz, the simplest solution is the LC L-section, illustrated in Figure 3.4.

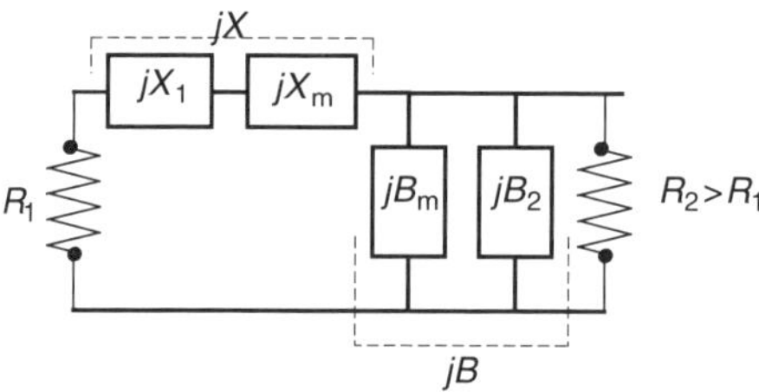

Figure 3.4: L-section matching set-up.

Two complex-valued impedances are shown, one in series form, the other in parallel form. The matching network consists of a series reactance, X_m, in line with the series-form impedance, and a susceptance, B_m, in shunt with the parallel-form impedance. The combined X and B values have to be such that R_1 is transformed to R_2. Think back to the tuned circuit of Figure 3.1. We saw that the small coil loss resistance was transformed to a high real resistance in parallel with the circuit. We can use the analysis equations in a design sequence. Given the terminating resistance values, the Q_r for the equivalent tuned circuit should be:

$$Q_r = \sqrt{\frac{R_2}{R_1} - 1} \tag{3.23}$$

The B and X values have to have the same sign. If you're choosing $X > 0$, the net series reactance is inductive and therefore $B > 0$ and the net shunt susceptance has to be capacitive. The remaining basic design equations are:

$$|X| = Q_r R_1, \quad |B| = Q_r / R_2 \tag{3.24}$$

What you do next depends on the particulars of the problem.

Example 3.4 **L-Section Matching**

To keep the numbers simple, let's suppose we have a load that's 5Ω in series with 20 μH and we want to match it to 50Ω at ω = 1 Mrad/s. From (3.24), $Q_r = 3$. Since $Q_r R_1 = 15\Omega$, and $X_1 = 20\Omega$, I choose $X > 0$, $X = 15\Omega$. Then $X_m = X - X_1 = -5\Omega$. Since I chose $X > 0$, $B > 0$, $B = B_m = 3/50 = 0.06$ S. Both matching elements are capacitors. $C_{\text{series}} = -1/(\omega X_{\text{m}}) = 0.2$ μF, $C_{\text{shunt}} = B_m/\omega = 0.06$ μF. The loaded Q of the series R-L load is 2, so its matched bandwidth should be 0.5. The matched bandwidth of the load and matching network turns out to be 0.545. It often happens in general, and almost always with small antennas, that the Q of the matching network is lower than that of the highest-Q termination.

3.2.2 Wideband Matching

In the previous section, the match was perfect at the design frequency. It is possible to trade perfection at a point for a pretty good match over a wider range of frequencies. This is what is meant by "wideband matching." For resistive terminations, analytical procedures are available for the network element design. If the load is complex, a numerical procedure is needed. Optimization is another area with a very large literature. The software CD-ROM accompanying this book has a set of programs to be run under MATLAB, which use the built-in search function fminsearch.m. The function can't tell the difference between a local minimum of the penalty function, and a global minimum, so the initial condition choice strongly influences the result. The supervisory program allows you to make either an educated or arbitrary guess at the initial conditions, and then steps through a loop in which an incremented fractional multiple of the initial condition elements is given to the optimizer on each pass. The optimizer results are saved in an array, and when the loop is finished the supervisor program finds the best result and the associated element values. Details of use are in the ice.m and ematch.m file comments on the CD-ROM.

Example 3.5 **Matching the Series-Tuned Loop**

Returning again to our loop antenna of Example 2.3, we know that its matched fractional bandwidth is about 0.008, or 0.008 × 30 MHz = 240 kHz. From Example 3.3, we know that parallel-resonating it gives a resistance of over 135 kΩ but series resonating it will give a resistance of about 6.06Ω. Since 50/6.06 << 135k/50, it makes better sense to series tune the loop and then match it to 50Ω. Series-tuning requires essentially the same capacitance as parallel-tuning,

5.859 pF. An L-section match calculation gives a series $L = 86.55$ nH and a shunt $C = 285.78$ pF. This gives a perfect match at 30 MHz and a half-power bandwidth of 234 kHz.

I decided to try optimizing over a 300-kHz bandwidth, from 29.85 to 30.15 MHz. The optimizer is set up to vary the element values in order to maximize the minimum gain in the specified frequency set. You should always have more matching frequencies than unknowns in the problem. If the number of match frequencies is equal to or less than the number of unknowns, you run the risk of finding a "perfect" solution which behaves very badly in between your specified frequencies. I used 10-kHz steps for a total of 31. The matching network has to have a lowpass ladder form—that is, alternating series L and shunt C elements. I tried ladders from 2 to 6 elements. With either lumped or distributed-element networks, there are many local minima, so the result depends strongly on the initial values for the search. I use a loop around the optimizer to step through a set of initial conditions, which improves the chance of finding the global minimum, but doesn't guarantee it. While the 5-element ladder gave a 0.15-dB improvement over the two-element case, I didn't think it was enough to warrant the extra three elements, so I'm giving just the two-element result. This is another L section with $C = 149.8$ pF and $L = 127.4$ nH. The half-power bandwidth is 329.5 kHz. The frequency responses are shown in Figure 3.5 and the schematic is shown in Figure 3.6 for both designs.

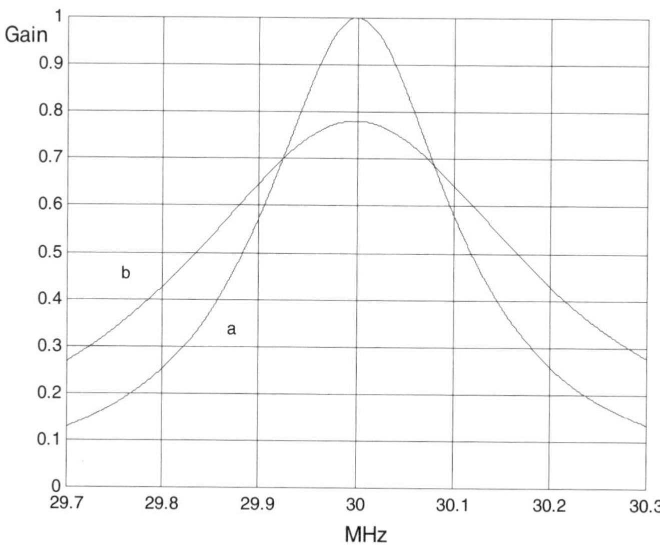

Figure 3.5: Matching network gain vs. frequency. (a) Single-frequency L section design. (b) Optimized L-section design.

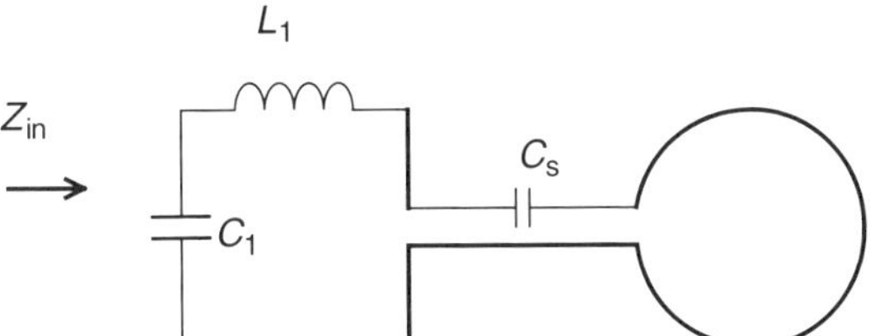

Figure 3.6: Matching network schematic. C_s = 5.8588 pF. For the single-frequency L section, C_1 = 286 pF, L_1 = 86.55 nH. For the optimized L section, C_1=149.8 pF, L_1 = 127.4 nH.

3.2.3 System Efficiency

As we saw in the dipole and loop examples, 2.2 and 2.3, the efficiency of the antenna by itself is not the most important part of system efficiency. By system, I mean both the antenna and the impedance-matching network necessary to interface it with the transmitter or receiver. For the dipole of Example 2.2, the reactance is –1307Ω. A resonating coil with a Q of 200 would have a loss resistance of 1307/200 = 6.535Ω. Because the resistances are all in series, the system efficiency would be:

$$Eff = R_{rad} \big/ \left(R_{rad} + R_{loss} + R_{coil}\right) = 1.739 \big/ \left(1.739 + 0.01887 + 6.535\right) = 0.2097$$

or 21%.

The efficiencies of the L-section-matched loop designs of Example 3.5 can be simply figured in almost the same way because the coil is directly in series with the loop. The difference is that the net loop resistance, R_{ext}, of 6.057Ω has to be proportioned in the same way as the internal R_{rad} and R_{loss} of the intrinsic loop. Call the total internal resistance R_{int}, and the external equivalents to $R_{rad,}$ R_{radx}. Then

$$R_{radx} = \frac{R_{rad} R_{ext}}{R_{int}} = \frac{1.923 \times 6.057}{2.1} = 5.546\Omega$$

Now we can calculate the coil resistances and efficiencies for the two matching designs. For the single-frequency design, $R_{coil} = \omega_o L/Q$ = 188.5M × 0.08655μ/200 = 0.0816Ω. The efficiency is 5.546/(6.057 + 0.0816) = 0.903 or 90.3%. For the optimized L section, R_{coil} = 188.5M × 0.1274μ/200 = 0.12Ω. The efficiency is 5.546/(6.057 + 0.12) = 0.898 or 89.8%. This is not the whole story though, because the insertion gain of the optimized L section at 30 MHz is just about 0.8. This reduces the system efficiency to about 70 % at the center frequency. The price of increased bandwidth is less power out at the center frequency.

You may be wondering why you see so many short dipoles around if they're so inefficient to match. Or you may be thinking, in view of Chapter 1, that I'm using a terrible case as an example. In the last section of this chapter I present some simple methods for improving the dipole, its monopole equivalent, and the loop. The dipole catches up pretty well, but you are right to be suspicious about all the little whips you see, because many of them are both short and unmatched and perform poorly compared to what is possible.

3.3 Reception

So far, I've been presenting antennas in transmit mode. How do they work in reception mode? We can learn a lot from two simple facts.

1. Faraday's Law. In the early 19th century, Michael Faraday experimentally determined that a time-varying magnetic field produces a voltage in a wire loop. If the loop is electrically small, the voltage is the time derivative of the magnetic force field perpendicular to the loop area, times the loop area. In phasor notation this is

$$V_{\text{loop}} = -j\omega B_n A = -j\omega\mu H_n A = -j\beta A E_t \tag{3.25}$$

   where the subscript "n" stands for "normal" (in the sense of perpendicular) and the subscript "t" stands for tangential. This gives us the open-circuit voltage source in the small loop's equivalent circuit. It also tells us that the maximum voltage will occur when the wave is oriented so that the **E**-field and the wave velocity vectors are in the plane of the loop.

2. Two antennas form a two-port network. Therefore, they can be represented by any general two-port equivalent circuit. Figure 3.7 shows an equivalent T network. In the following developments, all signal values are rms, so that we don't have unnecessary factors of 2 in the power expressions.

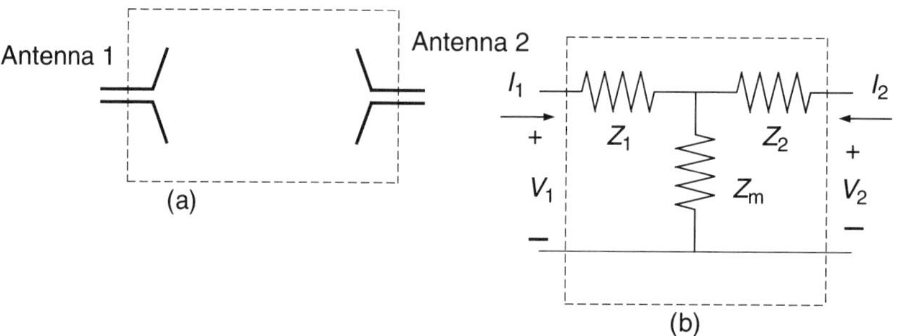

Figure 3.7: Two antennas as an equivalent T network.

In general, receiving antennas are characterized by either of two equivalent parameters, the *effective height*, h_e, or the *effective area*, A_e. These go with different visualizations of how the antenna works as a transducer. In the first way, one can imagine the antenna as a long wire intercepting the $\bar{E}$ field and converting it to voltage. The effective height is a number that multiplies $\bar{E}$ to give the open-circuit voltage. In the second picture, the antenna is a scoop, catching the power in the passing wave and converting it into power into a receiver. The effective area is a number that multiplies the power density in the passing wave to give the power into the load. In both cases the parameters are defined for best operating conditions. The wave is oriented and polarized to give the maximum antenna voltage, and the receiver load is matched to the antenna's impedance. These relations are symbolized by:

$$V_w = h_e E_{\text{inc}},\ P_L = A_e S_{\text{inc}} \tag{3.26}$$

3.3.1 Effective Height

In Figure 3.7, let Antenna 1 be an arbitrary antenna and Antenna 2 be a small loop. Assume that Antenna 1 is linearly polarized and the two antennas are oriented to produce the maximum coupling. Further, we are only interested in the far-field situation, so we may assume that $|Z_m| << |Z_1|, |Z_2|$. Consider two cases:

Case 1: Antenna 1 is transmitting, with input current I_1 and the loop is open-circuit with $V_{oc} = V_2$. In this case $V_{oc} = Z_m I_1$. Since Antenna 2 is a loop, we also know that $V_{oc} = -j\beta A E_{\text{inc}}$. The incident electric field magnitude is given by the radiation power density from Antenna 1.

$$\begin{aligned} \frac{|E_{\text{inc}}|^2}{\eta} &= I_1^2 R_{r1} \frac{D_1}{4\pi r^2}, \\ |E_{\text{inc}}| &= I_1 \sqrt{\eta R_{r1} D_1 / 4\pi r^2} \end{aligned} \tag{3.27}$$

with R_{r1} the radiation resistance and D_1 the directivity of Antenna 1. We can say then

$$|Z_m| = \beta A \sqrt{\eta R_{r1} D_1 / 4\pi r^2} \tag{3.28}$$

Case 2: The loop is driven with current I_2 and Antenna 1 is open circuit so $V_{oc} = V_1$. Now $V_{oc} = Z_m I_2$. By definition, $V_{oc} = h_{e1}|E_{\text{inc}}| = h_{e1}\eta\beta^2 A I_2/(4\pi r)$. So

$$|Z_m| = h_{e1}\eta\beta^2 A/(4\pi r). \tag{3.29}$$

Between these two results for $|Z_m|$ we can relate the receiving parameter to the transmit properties of Antenna 1.

$$h_{e1} = \lambda\sqrt{R_{r1}D_1 \big/ \eta\pi} \tag{3.30}$$

3.3.2 Effective Area

Since we now have an expression for the effective height, we can find the effective area with a little circuit analysis. Suppose we have an arbitrary lossless antenna in receiving mode with a matched load. The real part of the load impedance is R_r. The load current is $I = V_{oc}/(2R_r)$, so the load power is $P_L = I^2 R_r = V_{oc}^2/(4R_r) = (h_e^2 E^2)/(4R_r)$. But, by definition, $P_L = A_e S = A_e E^2/\eta$. Therefore,

$$A_e = \frac{\eta}{4R_r} h_e^2 = \frac{\lambda^2}{4\pi} D \tag{3.31}$$

This is an interesting result because the effective area doesn't seem to depend on the physical size directly. For our linearly polarized electrically small dipoles and loops, $D = 1.5$, so the effective area is indeed independent of the size. For larger antennas, the value of D increases with size in wavelengths, so there is an indirect connection to the physical size.

3.3.3 Reception Pattern

Suppose now that Antenna 1 has been turned away from the best direction (θ_m, ϕ_m) to some angles (θ, ϕ). Then D_1 in equation (3.27) has to be multiplied by the power pattern function, $f_p(\theta, \phi)$. Call the reception power pattern function $f_{pr}(\theta, \phi)$. Then in Case 2, section 3.3.1, the open-circuit voltage is multiplied by $\sqrt{f_{pr}(\theta, \phi)}$. These two changes cause (3.30) to read

$$\sqrt{f_{pr}(\theta, \phi)}\, h_{e1} = \lambda\sqrt{R_{r1}D_1 f_p(\theta, \phi) \big/ \eta\pi}$$

from which we may conclude that $f_{pr} = f_p$. The reception pattern is the same as the transmit pattern.

3.4 Ground Effects

We now leave the free space of imagination and begin to consider antennas in a space with other objects and materials. We saw in the last section that a transmitting antenna will induce current in a receiving antenna. This is a deliberate transfer of signal, but the radiation from the transmit antenna also induces currents in every conducting body it passes, and each of these induced currents re-radiates waves. This sounds like every object in the universe is making waves, but, while this is true in principle, as the objects become more distant from the transmit antenna, both the incident wave amplitude and the induced currents and re-radiation become vanishingly small. Large nearby objects have two major effects: they change the total radiation pattern, and they change the impedance of the transmit antenna by inducing currents back into the transmit antenna itself. The effect on the total radiation pattern can be thought of as a far-field effect, while the impedance change is a near-field effect.

In this section, we largely study the effects on the antenna of an infinite perfectly conducting plane. We do this partly because it actually approximates some practical situations, and partly because it is a simple enough problem to admit of some analysis so that we can observe effects that tend to occur in more complicated situations.

3.4.1 Image Theory

Image theory is the process of replacing a conducting plane, or group of planes, with an equivalent system of charges. We deal with only one plane. A perfect conductor enforces the condition that the tangential $\overline{\mathrm{E}}$ vector must be zero at its surface, so the $\overline{\mathrm{E}}$ field must be perpendicular at the surface of the plane. A single point charge above the plane will induce charges in the surface that counteract the tangential component of the point charge's field. The plane can be replaced by a single point charge of opposite sign, spaced as far below the plane as the original charge is above it. The total field of the two-charge system will meet the requirement of being perpendicular to the plane because their tangential components at the plane are in opposite directions and of equal magnitude. This is illustrated in Figure 3.8. The two-charge system in Figure 3.8(b) is equivalent to the charge-and-conductor system in Figure 3.8(a) in the sense that it meets the boundary condition and the total field above the plane is the same in both systems. However, in 3.8(a), there is no field below the plane, and this is an important point.

Currents are charges in motion, of course, and they have images too. We can picture how the images work if we realize that a current is moving positive charge from one end of a wire to the other, causing a "plus" charge accumulation in the direction the current is going, leaving a "minus" charge accumulation behind. These accumulations will have their opposite-sign images below the plane, and these will tell us where the image currents are going. This is illustrated in Figure 3.9.

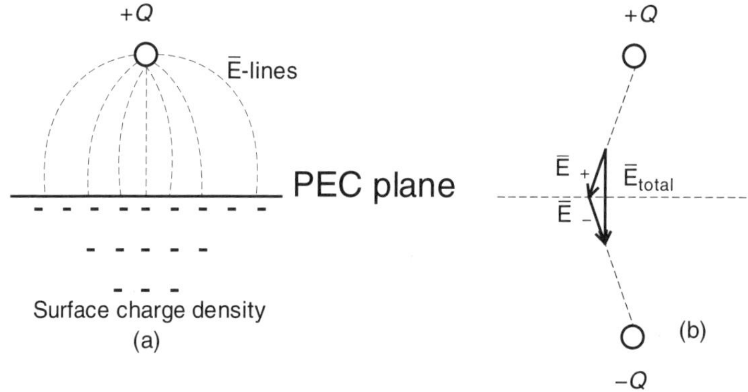

Figure 3.8: Illustration of basic image theory. (a) A point charge above a perfect electrically conducting (PEC) plane induces a negative surface charge density in the plane. The lines of electrical force are normal to the plane. (b) Equivalent image system replaces the plane with a negative charge to produce the same normal field where the plane was.

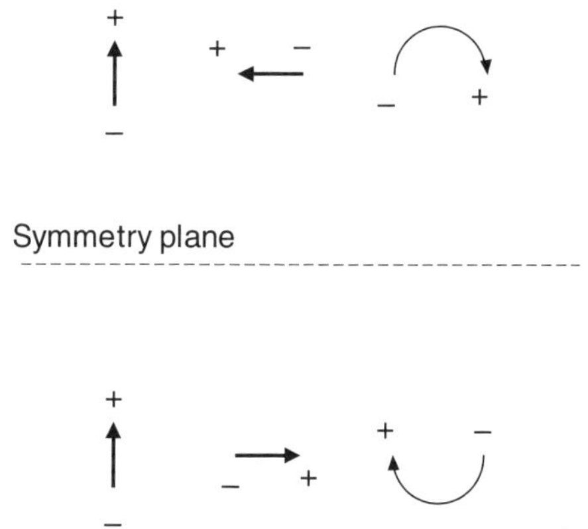

Figure 3.9: Current elements and their images.

3.4.2 Vertical Dipole Above a Perfect Ground Plane

First consider a short vertical dipole a distance a above a perfectly conducting plane. Let the coordinate origin be on the plane and the z axis go through the dipole. From Appendix B, we know that the radiated fields of the dipole will have a phase advance with respect to the origin. From Figure 3.9, the dipole will have an image below the symmetry plane at $z = -a$ which has the same fields, except for a phase delay. From equations (2.27) and (B.3–4), with $\rho' = 0$ and $z' = a$, the total field above the plane is

$$\begin{aligned} E_\theta &= \frac{j\eta L}{2\lambda}\alpha I_o \sin(\theta)\frac{e^{-j\beta r}}{r}\left(e^{j\beta a\cos(\theta)} + e^{-j\beta a\cos(\theta)}\right), \quad \alpha = \frac{I_{\text{average}}}{I_o} \\ &= \frac{j\eta L}{\lambda}\alpha I_o \sin(\theta)\frac{e^{-j\beta r}}{r}\cos(\beta a\cos(\theta)) \end{aligned} \tag{3.32}$$

At any point on the plane, the distance is the same to the antenna and its image, so the two fields add perfectly to give twice the original. If $a > 0.05\lambda$, points above the plane will have significant phase differences, reducing the field strength, as expressed by the cosine term in (3.32). This interference sharpens the radiation pattern over that of the single dipole.

The power pattern function is:

$$f_p(\theta) = \sin^2(\theta)\cos^2(\beta a\cos(\theta)) \tag{3.33}$$

I am interested in two parameters, the directivity and the radiation resistance. To get at these, I need to integrate the pattern function over the space above the plane. For $\beta a < 0.1$ I can approximate the cosine term and write:

$$\begin{aligned} f_p(\theta) &\approx \sin^2(\theta)\left(1 - 0.5\beta^2 a^2\cos^2(\theta)\right)^2 \\ &\approx \sin^2(\theta)\left(1 - \beta^2 a^2\cos^2(\theta) + \left(\frac{\beta^2 a^2}{2}\right)^2\cos^4(\theta)\right) \end{aligned} \tag{3.34}$$

The half-space integral is

$$\begin{aligned} F_v &= 2\pi\int_0^{\pi/2}\sin^3(\theta)\cos^2(\beta a\cos(\theta))\,d\theta \\ &\approx 2\pi\int_0^{\pi/2}\sin^3(\theta)\left(1 - \beta^2 a^2\cos^2(\theta) + \left(\frac{\beta^2 a^2}{2}\right)^2\cos^4(\theta)\right)d\theta \\ &\approx 2\pi\left(\frac{2}{3} - \frac{\beta^2 a^2}{15} + \frac{\beta^4 a^4}{70}\right) \end{aligned} \tag{3.35}$$

The approximation overstates the exact value by about 4.3% at $a = 0.1\lambda$. A plot of a numerical evaluation of the exact integral is given in Figure 3.10.

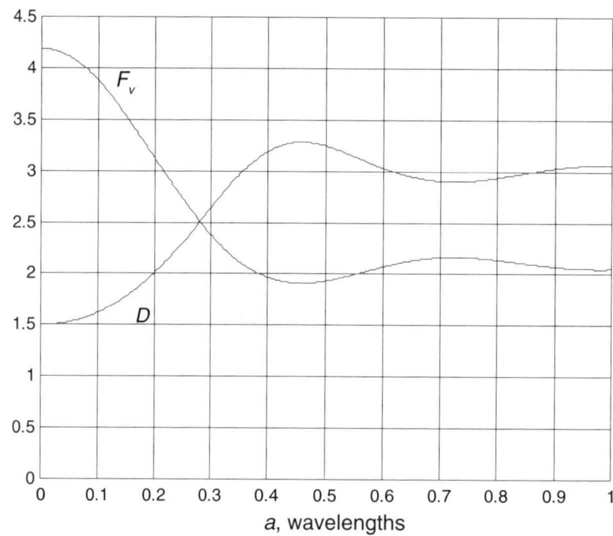

Figure 3.10: Plot of the power pattern integral F_v, and directivity *D*, for a vertical small dipole a distance *a* above a perfectly conducting ground plane.

Following the methods in section 2.3.2, the radiation resistance is:

$$R_{\text{rad}} = \eta\alpha^2 \left(\frac{L}{\lambda}\right)^2 F_v \tag{3.36}$$

For the open-wire short dipole, $\alpha = 1/2$, and when the antenna is close to the plane, $F_v = 4\pi/3$ from (3.35), so under these conditions,

$$R_{\text{rad}} = 40\pi^2 \left(\frac{L}{\lambda}\right)^2 \tag{3.37}$$

This is double the result found in section 2.3.2 for the short dipole in free space. Physically speaking, the antenna and its image produce twice the field strength, which translates to four times the power density of the isolated antenna. But the space radiated into is half the sphere, so we have four times the power density in half the area, giving a net twice the total radiated power. Twice the power for the same current means twice the resistance.

Directivity is an issue. Most people advertising their antenna over ground average the power density over the whole sphere, even though they are counting the radiation in the half-sphere. This gives them an automatic multiple of 2, or 3 dB. It is called gain referenced to an isotropic radiator in free space. The problem with this definition is that it gives a false idea about beamwidth and the general character of the radiation patterns. I define the directivity by averaging over the half-space, and this also makes a change in the expression for the maximum power density. Over a ground plane:

$$D = \frac{2\pi}{\int_0^{2\pi}\int_0^{\pi/2} f_p(\theta,\phi)\sin(\theta)d\theta d\phi} \tag{3.38}$$

$$S_{\max} = R_{\text{rad}} I_o^2 \frac{D}{2\pi r^2} \tag{3.39}$$

bearing in mind that the terminal current is rms in this section.

The directivity for the short vertical dipole is also plotted against altitude in Figure 3.10. It increases with altitude because more destructive interference (out-of-phase addition) occurs between the fields of the antenna and its image. This is the reason there is less total radiated power, and the resistance goes down.

3.4.3 Horizontal Dipole Above a PEC Plane

A horizontal dipole parallel to the x axis and a distance a above the plane has an image with opposing current, as seen in Figure 3.9. Following the same reasoning as with the vertical case, and using the field expression in (2.33), the total field above the plane is:

$$\bar{E} = \frac{-\eta L}{\lambda}\alpha I_o\left[\hat{\phi}\sin(\phi) - \hat{\theta}\cos(\phi)\cos(\theta)\right]\frac{e^{-j\beta r}}{r}\sin\left(\beta a\cos(\theta)\right) \tag{3.40}$$

Because any point on the plane is equally distant from the antenna and its image, their fields cancel and the plane is a null zone. The maximum field when $a \le \lambda/4$ is straight up, $\theta = 0$, and its amplitude is:

$$|E| = \frac{\eta L}{\lambda}\alpha I_o \frac{1}{r}\left|\sin(\beta a)\right| \tag{3.41}$$

The partial cancellation of the fields is represented by the $\sin(\beta a)$ term. When $a > \lambda/4$ the pattern splits and there are multiple maxima. This makes the normalization of the power pattern function a little different from what we've seen before.

$$\begin{aligned} f_p &= \left(\sin^2(\phi) + \cos^2(\phi)\cos^2(\theta)\right)\frac{\sin^2\left(\beta a\cos(\theta)\right)}{\sin^2(\beta a)}, \; a \le \lambda/4 \\ &= \left(\sin^2(\phi) + \cos^2(\phi)\cos^2(\theta)\right)\sin^2\left(\beta a\cos(\theta)\right), \; a > \lambda/4 \end{aligned} \tag{3.42}$$

Example 3.6 ***Field Plots for the Horizontal Dipole***

Suppose we have a dipole with $L = \lambda/10$, $\alpha = 1/2$, and $I_o = 1$A. Then the field magnitude at $r = 1$ km and $\phi = \pi/2$ is:

$$|E| = \frac{\eta L}{\lambda r}\alpha I_o\left|\sin(\beta a\cos(\theta))\right| = \frac{120\pi}{10\cdot 1000}0.5\left|\sin(\beta a\cos(\theta))\right| = 18.85\left|\sin(\beta a\cos(\theta))\right| \text{ mV/m}$$

This expression gives the field strength in a vertical plane perpendicular to the dipole.

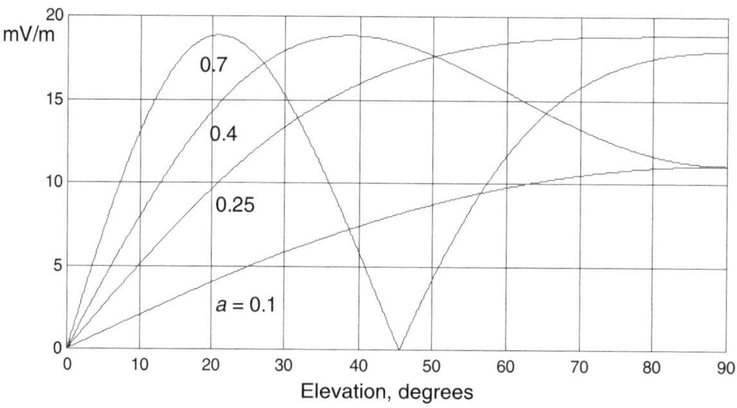

Figure 3.11: Field strength of the horizontal dipole plotted against elevation for various altitudes above a PEC plane (*a* is in wavelengths).

From Figure 3.11, you can see that when the altitude is below a quarter wavelength, the maximum field strength is correspondingly reduced from 18.85, and above a quarter wavelength is 18.85 but the location is shifted to a lower elevation. Also, as altitude is increased, more nulls and maxima appear. Mathematically, this is because the inner cosine goes through a quarter cycle, but the larger βa is, the more cycles the outer sine function is driven through. Physically, this is because the phase difference between the signals from the dipole and its image increases on the vertical line passing through them, and the projection of this phase difference on other lines passing through the origin becomes a multiple of 180° (addition) or 360° (cancellation).

The reduced maximum signal for altitudes under a quarter wave makes the expressions for the radiation resistance and directivity a little more complex because of the change in normalization factor. We have the integrated power pattern as:

$$F_h = \int_0^{2\pi}\int_0^{\pi/2}\left(\sin^2(\phi)+\cos^2(\phi)\cos^2(\theta)\right)\frac{\sin^2(\beta a\cos(\theta))}{\sin^2(\beta a)}\sin(\theta)\,d\theta d\phi,\ a \le \lambda/4$$
$$= \int_0^{2\pi}\int_0^{\pi/2}\left(\sin^2(\phi)+\cos^2(\phi)\cos^2(\theta)\right)\sin^2(\beta a\cos(\theta))\sin(\theta)\,d\theta d\phi,\ a > \lambda/4 \quad (3.43)$$

The radiation resistance is:

$$R_{\text{rad}} = \eta\left(\frac{L\alpha}{\lambda}\right)^2 \sin^2(\beta a)F_h,\ a \le \lambda/4$$
$$= \eta\left(\frac{L\alpha}{\lambda}\right)^2 F_h,\ a > \lambda/4 \quad (3.44)$$

The directivity expression is simple because the two-part nature is hidden in F_h:

$$D = \frac{2\pi}{F_h} \quad (3.45)$$

We can get a function that shows the altitude effect by normalizing the radiation resistance to $\eta(L\alpha/\lambda)^2$,

$$R_{\text{norm}} = \sin^2(\beta a)F_h,\quad a \le \lambda/4$$
$$= F_h,\quad a > \lambda/4 \quad (3.46)$$

In effect, R_{norm} is the second integral for F_h, over the whole range of βa. Numerical evaluation of F_h was used to generate the plots of D and R_{norm} shown in Figure 3.12.

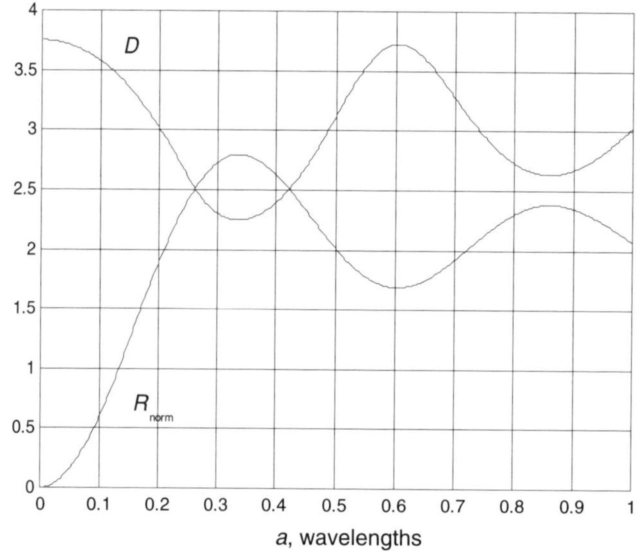

Figure 3.12: Plots of directivity and normalized radiation resistance against altitude for a horizontal dipole over a PEC plane.

You can see that the radiation resistance starts at zero at zero altitude, as expected. A perhaps surprising result is that the directivity peaks at zero altitude. This is foreshadowed in the $\overline{\text{E}}$-field curves, Figure 3.11, where you can see that the pattern for low altitude is simpler and sharper than those for higher altitudes.

If I had defined a normalized resistance for the vertical dipole case in the same way as for the horizontal case it would have been F_v. Comparing Figures 3.10 and 3.12, we can see that, as altitude increases, the values for F_v, D, and R_{norm} ripple around and close in to limiting values. The settling for the horizontal case is slower, but it appears that $D \rightarrow 3$, and F_v and $R_{\text{norm}} \rightarrow 2$. If you try some numerical examples, you will find that the unnormalized radiation resistance is quite close to the value for the dipole in free space.

3.4.4 Grounded-Source Antennas

If a ground plane is present, you don't have to put the antenna up in the air. You can drive one end of it and connect the ground of your cable to the plane. The image-theory picture of this situation is a little different than those seen in Figure 3.9, as shown in Figure 3.13.

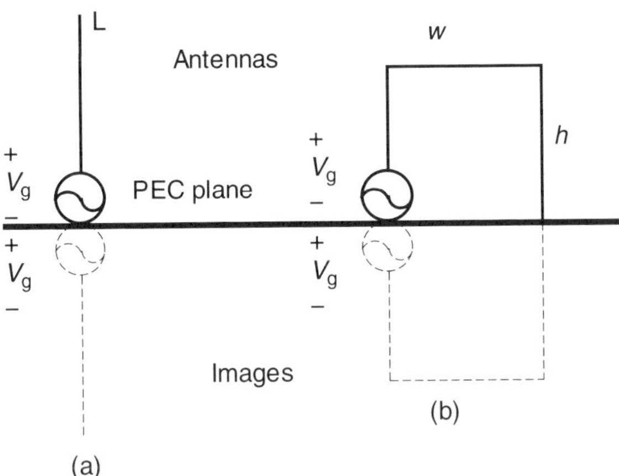

Figure 3.13: Grounded-source antennas and their images. (a) Monopole. (b) Half-loop.

In both cases shown, the antenna and its image make a new composite antenna with twice the height of the original. Also, the actual source voltage is half the applied voltage to the composite antenna. Consider first the monopole with length L. The composite antenna is a free-space dipole of length $2L$. The composite source sees an impedance appropriate to the length $2L$ and the wire diameter. The current through the actual source is $I = (2V_g)/Z_{\text{dipole}}$ so the impedance seen by the actual source is $Z_{\text{dipole}}/2$. Note that this result is a statement about the *total* impedance, not just the resistance as in the last two sections. For an electrically short monopole and composite dipole, the radiation resistance for a dipole of length $2L$ is four times that for a dipole of length L, but the monopole has half the resistance of the dipole of length $2L$, so the net result is that a monopole of length L has twice the radiation resistance of a free-space dipole of length L. By the same line of reasoning, the radiation resistance of the electrically small half-loop of area hw is twice that of a free-space loop of area hw.

3.4.5 Counterpoise

Monopoles for HF or lower frequencies are frequently placed over finite ground planes made up of wires. The wires can be in the form of a mesh, or laid out in radius lines from the monopole base. Traditionally, these artificial grounds have been either laid on top of the earth or buried a little distance. On the other hand, monopoles for frequencies from MF through UHF are placed on vehicles and the metal in the vehicle surface acts like a finite ground plane. Such a finite ground plane is also called a *counterpoise*. The idea of counterpoise has evolved to antennas in which there is a vertical radiator driven against a few horizontal elements (with canceling fields), the whole thing being elevated well off the earth.

Further, the antenna on a handheld radio may be driven against the radio box, which may be metal or metal-coated inside an outer plastic case. In this arrangement, the box is a counterpoise that also radiates. Some of these arrangements are presented in detail in Chapter 4.

3.4.6 Summary of Ground Effects

We have seen that the ground plane gives a performance boost to close dipoles that are perpendicular to the plane. On the other hand, dipoles that are parallel to the plane tend to be diminished in terms of their ability to radiate. Although I've not presented the analysis, these statements are true of loops close to the plane as well. These general behaviors are seen in real ground-proximity cases. However, in real grounds, or finite bodies that can be so approximated, the waves generated by the antenna penetrate the body and so induce currents and losses. The losses are reflected back to the driving source by means of an increased resistance. This resistance increases the bandwidth, but at the expense of decreased efficiency. If plenty of power is available on the transmit side of a link, then the lower efficiency may be acceptable. Otherwise, every effort should be made to minimize coupling to the nearby conductive object, whether it is the earth, in the case of HF and lower frequency antennas, or a human body, as in the case of handheld radio-telephones.

In some cases, the antenna on the handset is not the primary radiator, but is used to induce currents in the user's/wearer's body, which then are mainly responsible for the radiation [16]. The efficiency of such a system is low, but better than the tiny antenna by itself.

Currents are induced in all conducting objects, which then re-radiate waves. This fact means that the intended environment has to be considered when designing or using an antenna system. As a user of a radio handset, all sorts of domestic equipment, from your furnace to your refrigerator, will affect your radiation pattern and the antenna's impedance, for better or worse. In many cases, the user's best strategy is to move around to get the best signal possible. A major design strategy is to use two antennas with different polarizations to avoid a signal minimum. This latter strategy is discussed in a later chapter.

3.5 Improvements

The main problem with the small dipole is its low capacitance, causing the high series reactance. This problem has been attacked from two directions, (a) by making the wire thicker, and (b) by placing radial wires, or plates for physically small dipoles, at the ends to make it look more like a parallel-plate capacitor. The imitation-capacitor method has the additional advantage of increasing the average current on the radiating wire. This occurs because the current doesn't go to zero at the radiator ends, but spreads out on the imitation capacitor plates and goes to zero at the edges. In the limit, this effect yields a four-fold increase in the radiation resistance.

For the small loop, again making the wire or tubing thicker reduces the inductance. Also, using two loops in parallel with significant space between them reduces the inductance. Making a loop of a wide strap gives lower reactance.

The measures described in the last two paragraphs give lower Q but produce antennas that occupy more volume. In general, if size is limited in one or two dimensions, performance can be improved by using more of the remaining space. This idea was first published in 1947 by H. A. Wheeler [17]. A recent study by S. R. Best [18] using numerical simulation and experiment showed that occupied volume is still the principal determinant of antenna Q for three different types of small resonant monopoles.

If a small antenna can be made resonant or near-resonant at the operating frequency, then the impedance-matching part of the system can be designed with lower loss. There are two kinds of resonance, *wave resonance,* and *circuit resonance.* Wave resonance occurs because the current distribution on the structure is a multiple of a quarter-wave. This can be achieved by packing a lot of wire in a small space, as in a cylindrical spiral. Circuit resonance is achieved by building enough capacitance and inductance into the structure to achieve series resonance. Specific examples of small antennas exhibiting these resonances are given in later chapters.

References

[1] Frederick Terman, *Radio Engineers' Handbook*, McGraw-Hill, 1943.

[2] L. J. Chu, "Physical Limitations of Omni-Directional Antennas," J. Applied Physics, vol. 19, pp. 1163–1175, December, 1948.

[3] R. C. Hansen, "Fundamental Limitations in Antennas," Proc. IEEE, vol. 69, no. 2, pp. 170 182, Fcbruary, 1981.

[4] Wen Geyi, "Physical Limitations of Antenna," IEEE Trans. on Antennas and Propagation, vol. 51, no. 8, August 2003.

[5] Wen Geyi, "A Method for Evaluation of Small Antenna Q," IEEE Trans. on Antennas and Propagation, vol. 51, no. 8, p. 2127, August, 2003.

[6] A. D. Yaghjian and S. R. Best, "Impedance, Bandwidth, and Q of Antennas," Proc. IEEE Int'l. Symp. on Antennas and Propagation, Columbus, OH, 2003, vol. 1, pp. 501–504.

[7] H. L. Krauss, et. al. *Solid-State Radio Engineering*, Wiley, 1980.

[8] Matthaei, G. L., L. Young, and E. M. T. Jones, *Microwave Filters, Impedance Matching Networks, and Coupling Structures*, McGraw-Hill, 1964.

[9] Pozar, D. M., *Microwave Engineering*, Addison-Wesley, 1990.

[10] Miron, D. B., "Short-Line Impedance Matching; Some Exact Results," *RF Design*, March, 1992.

[11] Miron, D. B., "The LC Immittance Inverter," *RF Design*, Jan. 2000.

[12] Miron, D. B., "The Short-Line Transformer," *Applied Microwave and Wireless*, March 2001.

[13] Chen, Wai-Kai, *Passive and Active Filters: Theory and Implementation*, Wiley, 1986.

[14] Carlin, H. J., and P.P. Civalleri, *Wideband Circuit Design*, CRC Press, 1998.

[15] Miron, D. B., "Minimum-Length Cascades of Short Transmission Lines," *RF Design*, November 2003, p. 48ff.

[16] Andersen, J. B. and F. Hansen, "Antennas for VHF/UHF Personal Radio: A Theoretical and Experimental Study of Characteristics and Performance," IEEE Trans. on Vehicular Tech., vol. VT-26, no. 4, pp. 349–358, November, 1977.

[17] Wheeler, H. A., "Fundamental Limitations of Small Antennas," Proc. IRE, vol. 35, no. 12, pp. 1479–1484, December, 1947.

[18] Best, S. R., "A Discussion on the Quality Factor of Impedance-Matched Electrically Small Wire Antennas," IEEE Trans. on AP, vol. 53, no. 1, pp. 502–508, January 2005.

[19] Friis, H. T., "A Note on a Simple Transmission Formula," Proc. IRE, vol. 34, pp. 254–256, 1946.

Chapter 3 Problems

Section 3.1

3.1 Derive equation (3.5).

3.2 Derive equation (3.7).

3.3 Derive equations (3.9) and (3.10).

3.4 Suppose you need to design an RF amplifier to tune across the AM broadcast band, 550 to 1750 kHz. If a variable tuning capacitor with $C_{max} = 330$ pF is to be used, and a coil designed with a Q of 200 at 1750 kHz, find:

(a) The coil inductance needed to tune to 550 kHz,

(b) The minimum capacitance, C_{min}, needed to tune the coil to 1750 kHz,

(c) The coil loss resistance at 1750 kHz,

(d) The coil loss resistance at 550 kHz, assuming a skin effect resistance variation,

(e) The no-load bandwidth when the circuit is tuned to 550 kHz,

(f) The no-load bandwidth when the circuit is tuned to 1750 kHz.

3.5 Given the frequency dependence of coil reactance, and the frequency dependence of wire resistance, write an expression for the frequency variation of coil Q with respect to a known value at a known frequency. That is, find $f(\omega,\omega_o)$ such that $Q(\omega)=f(\omega,\omega_o)Q_o$.

3.6 Find analytical expressions for the derivatives needed in Example 3.1. Use them to find the bandwidth for SWR = 2.

3.7 Find analytical expressions for the derivatives needed in Example 3.2. Use them to find the half-power matched bandwidth.

Section 3.2

3.8 A variation on the classic maximum-power-transfer problem is shown in Figure P3.8. An ideal n:1 transformer, n not necessarily an integer, is connected between a generator and a load. R_x is a loss resistance. Find n to maximize the power in R_L.

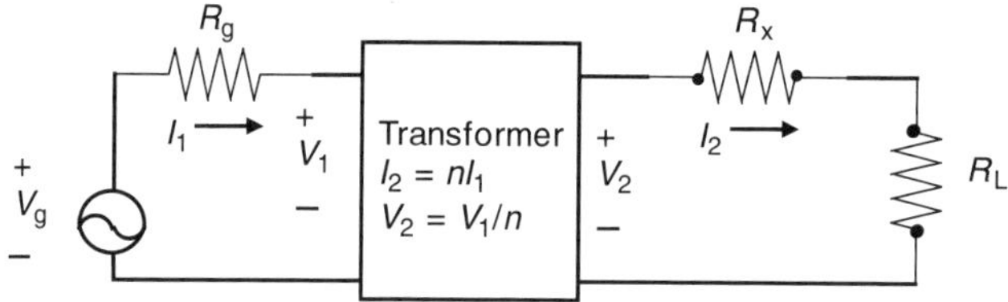

Figure P3.8: An idealized impedance-matching circuit.

Section 3.2.1

3.9 Design an L-section matching network for the loop of Example 2.3 at 20 MHz.

Section 3.2.2

3.10 Rework Example 3.5 for a 500-kHz bandwidth. That is, choose the matching frequencies as *fm* = [29.75:0.01:30.25]. Try various initial conditions and ladder lengths out to 6 elements.

3.11 Redesign C_s in Figure 3.6 to eliminate L_1 for the two L section designs. Plot the insertion gain of the two matching networks and compare it with Figure 3.5.

Section 3.3

3.12 Find the effective height and effective area for the small dipole of length L.

3.13 Find the effective height and effective area of the small loop.

3.14 Show that the maximum power delivered to the receiver by an antenna when antenna loss is taken into account is:

$$P_{\text{rec}} = \left(\frac{V_w}{2}\right)^2 \Big/ \left(R_{\text{loss}} + R_{\text{rad}}\right) = \left(\frac{V_w}{2}\right)^2 \frac{eff}{R_{\text{rad}}} = P_{\text{rec max}} eff \quad \textbf{(3.47)}$$

where P_{rec} is the receiver power under matched conditions, and P_{recmax} is what the matched receiver power would be if the receive antenna were lossless.

3.15 In view of the result in problem 3.14, show that the ratio of power into the receiver to power from the transmitter is:

$$\frac{P_{\text{rec}}}{P_{\text{tran}}} = G_t G_r \left(\frac{\lambda}{4\pi r} \right)^2 \quad (3.48)$$

where G_t is the transmitting antenna's gain, G_r is the receiving antenna's gain, and r is the distance between them. This is a version of the Friis Transmission Formula[19]. This equation applies in free space. In practice, on terrestrial radio links especially, one has to take into account additional wave attenuation caused by the atmosphere and earth. Notice that not only do the gains in this equation depend on the antenna sizes in wavelengths, but even the space between them is scaled in wavelengths.

3.16 Suppose we have a low-gain (1.5) antenna at both the receiver and transmitter sites, the receiver needs 10 μV into 50Ω, and the transmitter can supply 100W. At a 100-km distance, what is the highest frequency at which this link can be established? Use equation (3.48) in Problem 3.15.

3.17 Derive a version of (3.48) using effective areas instead of gains. Use the areas scaled in wavelengths squared.

Section 3.4.2

3.18 Derive the final approximate result in (3.35).

3.19 Derive the power pattern function for a vertical small loop at an altitude a above a PEC ground plane.

3.20 Write an expression for the radiation resistance of a small vertical loop just above a PEC ground plane.

3.21 Two antennas above a PEC plane are shown in Figure P3.21. Use the methods of section 3.3 and 3.4.2 to find an expression for the effective height of Antenna 1. You may assume that Antenna 1 is a small dipole and Antenna 2 is a small loop.

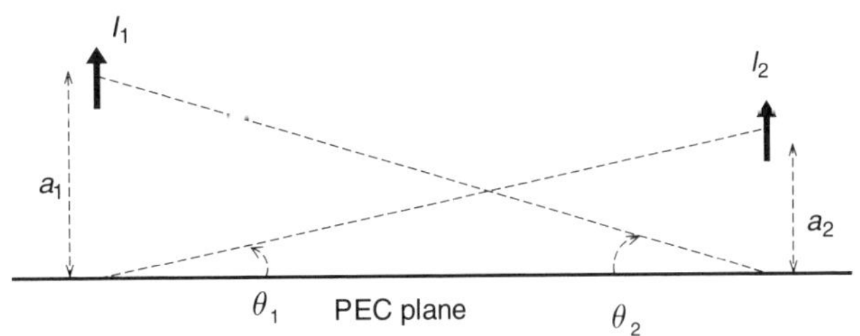

Figure P3.21: Two antennas over a perfectly conducting plane.

3.22 With respect to the discussion leading to (3.39), write an expression for the average radiation power density in terms of input power. Then write an expression for the maximum radiated power density in terms of input power.

Section 3.4.3

3.23 Derive an approximation for F_h, (3.43), when βa is small.

3.24 Calculate and plot the radiation resistances for both a vertical small dipole and a horizontal small dipole for altitudes of 2 to 10 wavelengths above a PEC plane. Choose $L = 0.1\lambda$, $\alpha = 1$. Comment on the results.

3.25 Consider a horizontal loop at an altitude a above a PEC plane.

(a) Sketch the loop and its image.

(b) Derive an expression for the radiated **E** field.

(c) Write the power pattern function.

(d) Calculate and plot the radiation resistance vs. altitude for $A = 0.01\lambda^2$.

Section 3.4.4

3.26 Calculate the input impedance of a half-loop over a PEC plane at 30 MHz. The half-loop is square, 1 m by 1 m, and has a tubing diameter of 25 mm. Neglect loss resistance.

Section 3.5

3.27 Consider a parallel-plate capacitor as an antenna. Let the plate area be A and the plate separation be h, and assume a radiating wire between them with a uniform current, $I_{\text{average}} = I_o$. Find an expression for the unloaded Q. This result sets a lower limit for what can be achieved by top-loading a monopole.

CHAPTER 4

Introduction to Numerical Modeling of Wire Antennas

4.1 General Concepts

Before the 1970s, engineers used mathematical analysis of rather simple problems and experimental experience as guides to design. The enormous growth of computer power since that time has added a third stage to the processes of research, development, and design, numerical simulation. Now the basic sequence of any of these processes is (1) analysis of relatively simple models, (2) numerical solution of the equations describing the full complexity of the target device or system, and (3) experimental verification. This statement assumes that we know the equations needed; if we don't, the problem is science, not engineering.

In stage (1), the models are approximate but the analytical solutions are exact. In stage (2), the models are as exact as we know them, and the numerical solutions are approximate. For example, if we want to design an audio amplifier to meet some requirements, and we choose to use a bipolar transistor of appropriate ratings, we can analytically evaluate a proposed design by modeling the transistor as a dependent current source between base and collector terminals and a resistor between the base and emitter terminals. The transistor's current gain will be assumed constant, and we may include or ignore internal capacitances. After writing and solving the frequency-domain circuit equations and choosing component values on this basis, we can go to the next stage. The next stage will use a very complex model of the transistor, which includes its internal capacitances, several more resistors and diodes, and parameters that depend on the terminal voltages and currents. This model is described by a set of nonlinear equations. These equations and the equations that describe the behavior and connections of the other amplifier components are solved in the time domain. At each point in time the unknowns and their derivatives are found by an *iterative method* (organized trial-and-error procedure) which gives an approximate solution to the current state and a projection as to where the state is going. A small time step is taken and the process repeated. There are various ways to improve the solution's accuracy and various mistakes that can lead to wildly inaccurate results.

Each voltage and current in a circuit problem is a function of a single variable, either time or frequency. In antennas, the essential problem is to find the current at each point on the structure. In Chapter 2 I assumed current distributions for the dipole and loop, and said they were approximate but useful for the electrically small versions of these antennas. For more complex antennas, there is no analytical guidance that justifies any assumption about the current distribution. Once we know the current distribution, we can calculate the far-field wave values and all the pattern parameters. So current is the basic unknown and is a function of position as well as either time or frequency. The methods presented in this text are in the frequency domain.

For a fixed frequency, the problem reduces to finding a continuous complex-valued function of position. A related problem is the spectral analysis of a periodic signal. In this case, we know a function of time over a particular interval and we want to know what the amplitudes of the component frequencies are. We know that the periodic function can be approximated by a series of harmonic sine and cosine functions, and we can use the Fourier series formulas to find the coefficients (amplitudes) of these sinusoids. The approximation gets better as one uses more and higher-frequency terms in the series. Likewise, the current on an antenna can be represented by a finite series of known functions with unknown amplitudes. Instead of adding up to a known function, though, the antenna current has to satisfy a boundary condition.

The current on a simple structure like a dipole can be represented as a Fourier series. Each function in the series is defined over the entire length of the dipole, so it is called an *entire-domain basis function*. More complex structures are modeled using current functions defined over only short parts of the structure. These are called *sub-domain basis functions*. An example using triangles is shown in Figure 4.1. Here, each triangle is nonzero from the center of its neighbor to the left to the center of its neighbor to the right. The function at a point is the sum of the values of the two (or one, at the ends) triangles that span the point. If the function is a current it would, in general, be complex-valued so each triangle coefficient would be complex-valued, and a sketch like Figure 4.1 would have a plot for the real part and another plot for the imaginary part. The accuracy of this representation improves when the domain of each basis function is shortened, requiring more functions and more unknown coefficients.

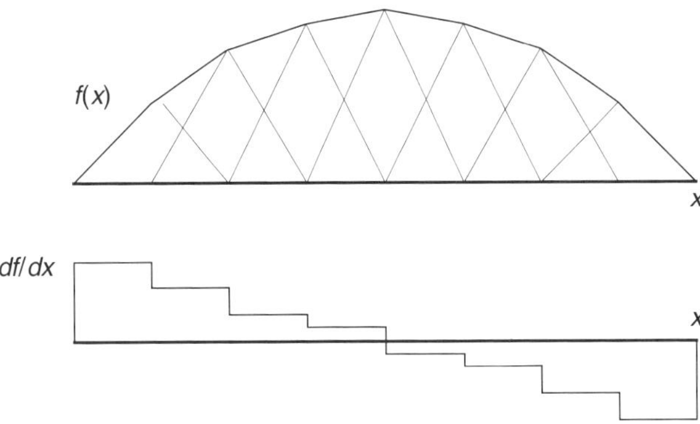

Figure 4.1: A piecewise-linear representation of *f*(*x*), using triangle sub-domain basis functions. *df*/*dx* is a sequence of piecewise-constant values.

4.2 The Mathematical Basics of the Numerical Electromagnetic Code (NEC)

NEC started out in the early 1970s as AMP (Antenna Modeling Program), which was developed for various military research offices. It was refined and further developed at Lawrence Livermore National Laboratory (LLNL), again with funding from various military offices, during the 1980s, and the current version, NEC4.1, was issued in 1992. NEC2 (from 1981) [1] is widely available as free software [2] in the public domain. NEC4.1 requires a license from LLNL, and its distribution is restricted. NEC is designed to solve problems involving round wires and surface patches. The following material concerns only the wire-modeling portion as this is more applicable to small structures. The presentation is intended to help you understand the reasons for the use guidelines described in later sections, so few of the actual formulas used in the code are given.

As mentioned in Chapter 3, the tangential electric field at a perfect conductor must be zero. Likewise, the normal magnetic field must also be zero at the surface of a perfect conductor. Just as the electric field of a wave can be written as an integral of the current over the structure, so the electric field at a point on the surface can be written as an integral of the current over the structure. The integral contains a term similar to the far-field term we saw in Chapter 2, plus a term which expresses the Coulomb force due to the charges equivalent to the varying current. In physical fact, both the current and the field are in the surface of the conductor, and the integrand has terms in powers of $1/R$, R being the distance between a current element and a field point. Enforcing the boundary condition where the field and source point are the same leads to an unbounded integrand. This singularity problem has been dealt with in many ways over the years. The approach for wires taken in NEC is called the thin-wire approximation. R is prevented from going to zero by saying the surface current is equivalent to a current concentrated on the wire's axis, and the field to be made zero is at the wire's surface. Figure 4.2 illustrates some of the ideas involved. Two arbitrarily located and oriented wire sections are shown. Here s_i is a variable that measures distance along wire i and it has two associated unit vectors, one on the wire axis where the current is concentrated, and the other on the wire surface where the field is calculated. In a general coordinate system, the field at a point $\bar{r}$ on wire i due to a current element at $\bar{r}'$ on wire k is

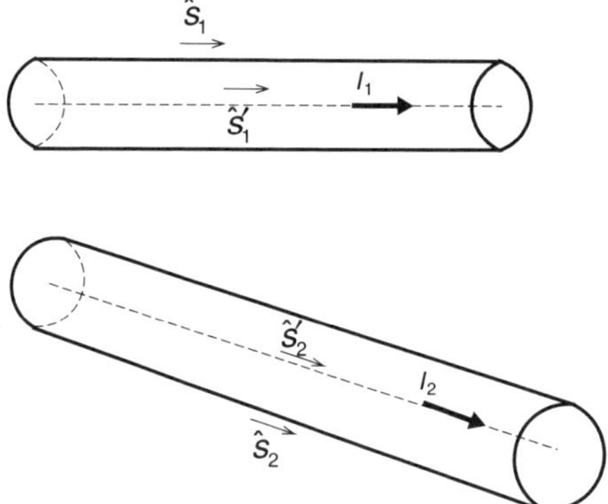

Figure 4.2: Two wire segments, illustrating equivalent current on centerline, source (primed) unit vectors, and surface tangential (unprimed) unit vectors.

$$d\bar{E}_i(\bar{r}) = \left[\frac{-j\omega\mu_o}{4\pi} I_k(\bar{r}')\hat{s}'_k g(R) - \frac{1}{4\pi j\omega \epsilon_o} I_k(\bar{r}')\hat{s}_i \frac{\partial^2 g(R)}{\partial s_i \partial s'_k} \right] ds'_k \tag{4.1}$$

with $g(R)$ being the propagation function from Chapter 2,

$$g(R) = \frac{e^{-j\beta R}}{R}, \quad R = |\bar{r} - \bar{r}'| \tag{4.2}$$

An antenna in transmit mode has an electric field applied at its feedpoint, and in receive mode there is an incident wave. In both cases, the externally applied electric field, $\bar{E}_a$, adds to the field of the wire currents. The tangential sum must be zero, and the tangential component is picked out of the sum by way of a unit vector dot product.

$$\begin{aligned}
\left(\bar{E}_a(\bar{r}) + d\bar{E}_i(\bar{r})\right)\cdot\hat{s}_i &= \bar{E}_a(\bar{r})\cdot\hat{s}_i + \left[\frac{-j\omega\mu_o}{4\pi} I_k(\bar{r}')\hat{s}'_k\cdot\hat{s}_i g(R) - \frac{1}{4\pi j\omega \epsilon_o} I_k(\bar{r}') \frac{\partial^2 g(R)}{\partial s_i \partial s'_k} \right] ds'_k \\
\bar{E}_a(\bar{r})\cdot\hat{s}_i &= \sum_k \int_0^{L_k} \left[\frac{j\omega\mu_o}{4\pi} I_k(\bar{r}')\hat{s}'_k\cdot\hat{s}_i g(R) + \frac{1}{4\pi j\omega \epsilon_o} I_k(\bar{r}') \frac{\partial^2 g(R)}{\partial s_i \partial s'_k} \right] ds'_k
\end{aligned} \tag{4.3}$$

Equation (4.3) expresses the current-generated field as the sum over all the wires of integrals over each wire. The wires don't have to be connected to fit into this formulation. To solve this equation for the current, we need a model for the applied field, and we need a set of basis functions with which to represent the current.

4.2.1 Basis Functions

The developers of NEC made a great effort to find basis functions that are smooth, that provide both current and charge continuity at junctions, and that can be substituted into (4.3) and the integration done analytically. The basis function they came up with has several parts in space and three component functions for each segment of wire for which it is defined. Consider first a three-segment piece of wire with segment numbers 1, 2, 3. Let s be the distance variable along the wire, s_k be the coordinate of the center of segment k, and d_k be the length of segment k. The basis function associated with segment 2 is $f_2(s)$ and it has three parts: $b(s)$ defined on segment 1, $m(s)$ defined on segment 2, and $a(s)$ defined on segment 3. I chose the letters as "b"efore, "m"ain", and "a"fter, to correspond to the assumed direction of positive current flow.

$$
\begin{aligned}
&f_2(s)=\{b(s),m(s),a(s)\} \\
&b(s)=A_b+B_b\sin(\beta(s-s_1))+C_b\cos(\beta(s-s_1)),\ |s-s_1|<d_1/2 \\
&m(s)=A_m+B_m\sin(\beta(s-s_2))+C_m\cos(\beta(s-s_2)),\ |s-s_2|<d_2/2 \\
&a(s)=A_a+B_a\sin(\beta(s-s_3))+C_a\cos(\beta(s-s_3)),\ |s-s_3|<d_3/2
\end{aligned}
\tag{4.4}
$$

This basis function has nine coefficients. On each segment there is a function component that is constant, one with odd symmetry about the segment center, and one with even symmetry about the segment center. There are smoothness conditions applied at each end of the basis function, $f_2(s) = 0$ and $df_2/ds = 0$ at the far ends of segments 1 and 3, and continuity conditions on the component functions and their derivatives at the segment junctions. Altogether, there are eight equations, so that one coefficient is free in the end. A final condition is applied to set the amplitude of the basis function. In NEC2, it is $A_m = -1$. Since the total amplitude of f_i at the midpoint of segment i is $A_m + C_m$, the condition on A_m alone makes the basis function amplitude dependent on modeling choices, particularly segment length. The consequences of this are discussed below.

To help our understanding of the conditions and behavior of these basis functions, I have written and solved the boundary equations for the general three-segment case, with all segments having the same wire diameter. To simplify the presentation, let each segment have its own coordinate which is zero at the middle, s_i' for segment i. Then at the left end, $s_i' = -d_i/2$ and at the right end $s_i' = +d_i/2$. This arrangement is shown in Figure 4.3, along with a plot of $f_2(s)$.

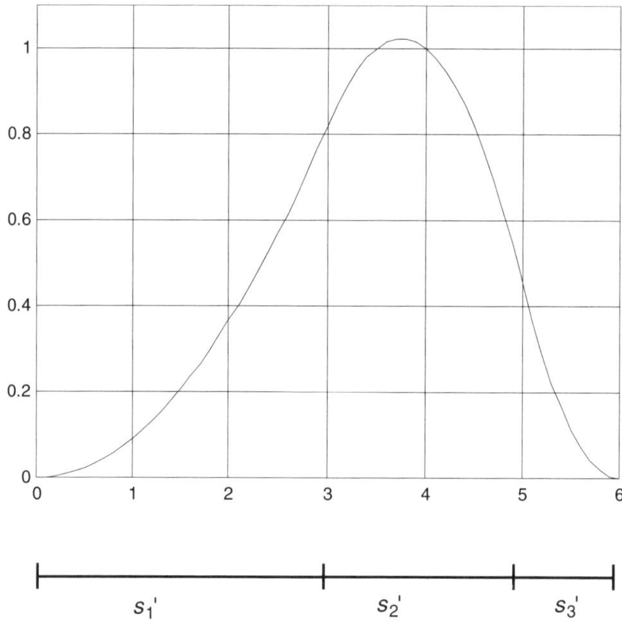

Figure 4.3: Illustration of a three-segment basis function. The segment lengths are d_1 = 3, d_2 = 2, and d_3 = 1. The primed coordinate is defined only for its associated segment, and is zero in the middle. λ = 60. This plot was done requiring $A_m + C_m = 1$. Notice that the uneven segment lengths cause the actual function peak to be a little larger than 1, and to the left of the center of segment 2.

Using the local coordinates, the basis function definition becomes:

$$
\begin{aligned}
&f_2(s) = \{b(s_1'), m(s_2'), a(s_3')\} \\
&b(s_1') = A_b + B_b \sin(\beta s_1') + C_b \cos(\beta s_1'), \quad -d_1/2 < s_1' < d_1/2 \\
&m(s_2') = A_m + B_m \sin(\beta s_2') + C_m \cos(\beta s_2'), \quad -d_2/2 < s_2' < d_2/2 \\
&a(s_3') = A_a + B_a \sin(\beta s_3') + C_a \cos(\beta s_3'), \quad -d_3/2 < s_3' < d_3/2
\end{aligned}
\tag{4.5}
$$

The line charge density ρ_l, in C/m, is related to the current's derivative by the continuity equation:

$$
\frac{dI}{ds} + j\omega\rho_l = 0 \tag{4.6}
$$

So, charge continuity at a junction requires continuity in the current's derivative. The derivatives of the basis function components are:

$$
\begin{aligned}
\frac{db}{ds} &= B_b\beta\cos(\beta s_1') - C_b\beta\sin(\beta s_1'), \quad -d_1/2 < s_1' < d_1/2 \\
\frac{dm}{ds} &= B_m\beta\cos(\beta s_2') - C_m\beta\sin(\beta s_2'), \quad -d_2/2 < s_2' < d_2/2 \\
\frac{da}{ds} &= B_a\beta\cos(\beta s_3') - C_a\beta\sin(\beta s_3'), \quad -d_3/2 < s_3' < d_3/2
\end{aligned}
\tag{4.7}
$$

At the left-hand end of segment 1, $f_2 = df_2/ds = 0$, which is $b = db/ds = 0$.

$$A_b - B_b\sin\left(\beta\frac{d_1}{2}\right) + C_b\cos\left(\beta\frac{d_1}{2}\right) = 0 \tag{4.8}$$

$$B_b\cos\left(\beta\frac{d_1}{2}\right) + C_b\sin\left(\beta\frac{d_1}{2}\right) = 0 \tag{4.9}$$

The two equations can be used to get expressions for B_b and C_b in terms of A_b. Likewise, at the right-hand end of segment 3, $f = df/ds = 0$ means $a = da/ds = 0$.

$$A_a + B_a\sin\left(\beta\frac{d_3}{2}\right) + C_a\cos\left(\beta\frac{d_3}{2}\right) = 0 \tag{4.10}$$

$$B_a\cos\left(\beta\frac{d_3}{2}\right) - C_a\sin\left(\beta\frac{d_3}{2}\right) = 0 \tag{4.11}$$

At the junction between segments 1 and 2, $b - m = 0$ and $db/ds - dm/ds = 0$.

$$A_b + B_b\sin\left(\beta\frac{d_1}{2}\right) + C_b\cos\left(\beta\frac{d_1}{2}\right) - A_m + B_m\sin\left(\beta\frac{d_2}{2}\right) - C_m\cos\left(\beta\frac{d_2}{2}\right) = 0 \tag{4.12}$$

$$B_b\cos\left(\beta\frac{d_1}{2}\right) - C_b\sin\left(\beta\frac{d_1}{2}\right) - B_m\cos\left(\beta\frac{d_2}{2}\right) - C_m\sin\left(\beta\frac{d_2}{2}\right) = 0 \tag{4.13}$$

At the junction between segments 2 and 3 $m - a = 0$ and $dm/ds - da/ds = 0$.

$$A_m + B_m\sin\left(\beta\frac{d_2}{2}\right) + C_m\cos\left(\beta\frac{d_2}{2}\right) - A_a + B_a\sin\left(\beta\frac{d_3}{2}\right) - C_a\cos\left(\beta\frac{d_3}{2}\right) = 0 \tag{4.14}$$

$$B_m\cos\left(\beta\frac{d_2}{2}\right) - C_m\sin\left(\beta\frac{d_2}{2}\right) - B_a\cos\left(\beta\frac{d_3}{2}\right) - C_a\sin\left(\beta\frac{d_3}{2}\right) = 0 \tag{4.15}$$

Now the amplitude-setting condition in NEC2 is:

$$A_m = -1 \tag{4.16}$$

I used

$$A_m + C_m = 1 \tag{4.17}$$

to investigate a simple way of keeping the basis-function amplitudes near 1, regardless of modeling choices. Solving (4.8–4.15) with (4.17) gives the following:

Define

$$R = \frac{A_b}{A_a} = \frac{\sin\left(\beta\frac{d_3}{2}\right)\sin\left(\beta\frac{d_3+d_2}{2}\right)}{\sin\left(\beta\frac{d_1}{2}\right)\sin\left(\beta\frac{d_1+d_2}{2}\right)} \tag{4.18}$$

$$C_m = \cos\left(\beta\frac{d_2}{2}\right)\Bigg/\left[\cos\left(\beta\frac{d_2}{2}\right) - 1 + \frac{4\sin\left(\beta\frac{d_2}{2}\right)\sin\left(\beta\frac{d_3}{2}\right)\sin\left(\beta\frac{d_3+d_2}{2}\right)}{R\sin(\beta d_1) + \sin(\beta d_3)}\right] \tag{4.19}$$

$$A_m = 1 - C_m \tag{4.20}$$

$$A_a = \frac{A_b}{R} = 2C_m\frac{\sin\left(\beta\frac{d_2}{2}\right)}{R\sin(\beta d_1) + \sin(\beta d_3)} \tag{4.21}$$

$$B_m = C_m\tan\left(\beta\frac{d_2}{2}\right)\left(1 - \frac{2\sin(\beta d_3)}{R\sin(\beta d_1) + \sin(\beta d_3)}\right) \tag{4.22}$$

$$B_a = -A_a\sin\left(\beta\frac{d_3}{2}\right) \tag{4.23}$$

$$C_a = -A_a\cos\left(\beta\frac{d_3}{2}\right) \tag{4.24}$$

$$B_b = A_b\sin\left(\beta\frac{d_1}{2}\right) \tag{4.25}$$

$$C_b = -A_b\cos\left(\beta\frac{d_1}{2}\right) \tag{4.26}$$

Observe that if $d_1 = d_3$, we have even segment-length symmetry and B_m, the main segment sine coefficient, is zero. The sine coefficients for the tail functions are never zero, because these functions have to have an odd symmetry component to get from a finite value at the junction with the main segment to zero at the far end. Since we are interested in small antennas, the next thing to do is examine the coefficient values for the case of short segments. Assume all the segment lengths are d and $d << \lambda$. The sine terms will reduce to their arguments, and the cosine terms will reduce either to 1 or $\cos(x) = 1-x^2/2$, whichever is more useful. This gives the following values:

$$C_m = 8\Big/\left[3(\beta d)^2\right] \tag{4.27}$$

$$A_a = A_b = \frac{C_m}{2} = 4\Big/\left[3(\beta d)^2\right] \tag{4.28}$$

$$B_b = -B_a = \frac{2}{3\beta d} \tag{4.29}$$

From (4.24) and (4.26), $C_{b,a} \approx -A_{b,a}$, but the mid-segment value of the tail function is:

$$A_{b,a} + C_{b,a} = \frac{(\beta d)^2}{8} A_{b,a} = 1/6 \tag{4.30}$$

As mentioned earlier, one of the motivations for using the three-term version of the component functions is that the field due to each term has been found analytically. The field at a given enforcement point due to a given component function will be found as the difference between two large numbers, something like finding the field due to a 1A current by subtracting the fields due to two 50,000A currents.

The general solution for the coefficients in NEC2, using $A_m = -1$ as the amplitude-setting condition, is given in [1, part I pp. 16–18]. Boiling these equations down to our short uniform three-segment case gives:

$$C_m = 1 + \frac{3}{8}(\beta d)^2 \tag{4.31}$$

$$A_{b,a} = 0.5, \quad C_{b,a} = -0.5, \quad A_{b,a} + C_{b,a} = \frac{(\beta d)^2}{16} \tag{4.32}$$

$$B_b = -B_a = \frac{\beta d}{4}. \tag{4.33}$$

The ratio of the mid-segment value of the tail functions to that of the main function is 1/6, just as in the previous case. These results show that the fields of 1A currents are subtracted to produce the fields of microampere currents, which are the matrix entries in the solution of the problem. Not only is there a problem with the small difference between close numbers, the fields at any distance from a source current are pushed toward the small-number limit of the computer/software system. Both of these problems are eased somewhat by using double-precision arithmetic as in the NEC2DSX*.exe compilations.

In NEC4.1 [3], two changes were made concerning the basis functions. The cosine was replaced by $\cos(\beta s_k') - 1$, which means the constant term is $A - C$, and the amplitude-setting condition is $A_m - C_m = -1$. Numerical solution of these equations for our short uniform three-segment basis function leads to:

$$C_m = 1 + \frac{3}{8}(\beta d)^2, \quad A_m = \frac{3}{8}(\beta d)^2 \tag{4.34}$$

$$A_{b,a} = \left(\frac{\beta d}{4}\right)^2, \quad B_b = -B_a = \frac{\beta d}{4}, \quad C_{b,a} = -0.5 \tag{4.35}$$

Because of the function change, C doesn't contribute to the mid-segment value of the component functions, which is now just A.

Example 4.1 **Fitting a Short Sine Function**

In Chapter 2 the analytical model for the current on a short dipole was given as $I(z) = I_o\sin(\beta(L/2 - |z|))/\sin(\beta L/2)$, with L being the dipole length, I_o the terminal current, and $z = 0$ at the dipole center. I have written a MATLAB function to generate this current distribution for an input data set. I have also written functions that set up and solve the equations for the basis function coefficients and then use the basis functions to fit mid-segment data points. All these functions are on the CD-ROM in the MATRF folder.

Figure 4.4 shows the data points for a 1-m dipole, $\lambda = 10$ m, divided into 10 segments, driven by 1A. You can see that the data fit is exact, which is always the case because the number of unknown coefficients is always equal to the number of data points. If a line for the dipole current had been plotted also, it would have been indistinguishable from the fitting function except between $s = 0.45$ and 0.55. The subplots are each a basis function multiplied by its fitting coefficient, so there are actually only three different kinds of basis functions used. Eight of the basis functions are the standard three-segment symmetrical function discussed above. The function at the left end has a main and a tail component, and the function at the right end has a tail and a main component. Notice that the interior basis functions all go to zero with zero slope as described earlier. At the wire ends, however, the basis functions go to zero with a nonzero slope. The slope is left free and that leaves five boundary conditions on the two segments, plus the amplitude-setting condition, to nail down the six coefficients in a two-segment basis function. This mathematical arrangement allows the end basis function to model the fact that the current has to go to zero at the open end of the wire, but the charge does not.

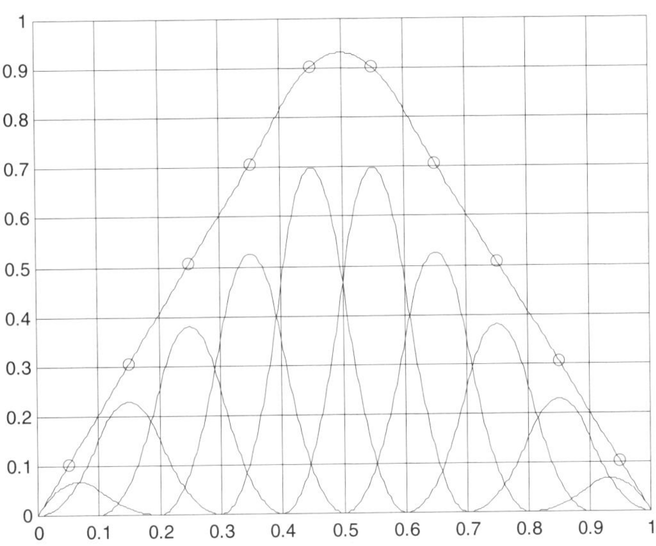

Figure 4.4: NEC basis functions used to fit a short-sine data set defined at the midpoints of ten segments. The circles mark the data points being fitted. The short-sine function has a value at (0.5,1) but this is not a data point because it is at a segment boundary. The wire length is $\lambda/10$.

The segment lengths are all the same, so $B_m = 0$ for the eight interior functions, and has the same magnitude for the two end functions. I have run the fitting for the NEC4, NEC2, and quasi-normalized ($A_m + C_m = 1$) NEC (QNEC) basis functions. Some of the component function coefficients and the fitting coefficients are listed in Table 4.1.

Table 4.1: Coefficients for the short-sine fit, $L/\lambda = 0.1$, $d/\lambda = 0.01$.

Interior Basis Function

	A_b	B_b	C_b	A_m	C_m
QNEC	337.88	10.613	−337.71	−674.42	675.42
NEC2	0.50099	0.015736	−0.50074	−1	1.0015
NEC4	0.000247	0.015736	−0.50074	0.0014828	1.0015

End Basis Function

	A_m	B_m	C_m	A_a	B_a	C_a
QNEC	−1214.8	12.736	1215.8	405.48	−12.736	−405.28
NEC2	−1	0.010484	1.00082	0.33377	−0.010484	−0.33361
NEC4	8.2e−4	0.010484	1.00082	1.65e−4	−0.010484	−0.33361

Fitting Coefficients, Segments 1–5

QNEC	0.06357	0.22839	0.38046	0.52602	0.69866
NEC2,4	77.226	154.03	256.59	354.75	471.19

The value for NEC4's A_m is also the overall basis function mid-segment amplitude in NEC2 and NEC4. Since this is quite small, it leads to the need for much larger fitting coefficients than those needed for the QNEC basis functions. I also ran the data and fitting programs for $\lambda = 100$. The plot looks essentially the same, but the coefficients changed dramatically, as shown in Table 4.2.

Table 4.2: Coefficients for the short-sine fit, $L/\lambda = 0.01$, $d/\lambda = 0.001$.

Interior Basis Function

	A_b	B_b	C_b	A_m	C_m
QNEC	33,774	106.1	–33,774	–67,546	67,547
NEC2	0.5	0.00157	–0.5	–1	1.0000148
NEC4	2.47e–6	0.00157	–0.5000074	14.8e–6	1.0000148

End Basis Function

	A_m	B_m	C_m	A_a	B_a	C_a
QNEC	–121,584	127.32	121,585	40,529	–127,324	–40,528
NEC2	–1	0.001047	1.0000082	0.3333377	–0.001047	–0.3333361
NEC4	8.2e–6	0.001047	1.0000082	1.645e–6	–0.0010472	–0.3333361

Fitting Coefficients, Segments 1–5

QNEC	0.06253	0.22492	0.3757	0.5213	0.697
NEC2,4	7,603	15,193	25,377	35,210	47,079

As pointed out in the NEC4 manual, if one is running single-precision arithmetic, there is a precision problem for small βd. In single precision you only have 6–7 significant figures, so if you are subtracting two fields that are the same for 4–5 significant figures, you don't have many significant figures left. The solution to this problem for NEC2 users is to always use double-precision versions, which use around 18 significant figures. As you can see in Tables 4.1 and 4.2, in NEC4 the A and C coefficients are quite different. In NEC4, the field due to a $\cos(x) - 1$ current function is calculated first and then multiplied by C. This means that C, a number near 1, is multiplied by a small number. The field due to a $\cos(x) - 1$ current is calculated partly analytically and partly numerically, and special approximations are used for short segments and close-in field points. While these measures may have helped in the case of single-precision arithmetic, I have not found cases in which NEC2 failed but NEC4 didn't, using double-precision versions.

Another thing that should concern the user is the shrinking of the basis-function amplitudes. This means that the field values that go into the current matrix equation (discussed later) shrink rapidly as the segment length. This effect can also lead to reduced accuracy in the final solution.

Example 4.2 Fitting a Constant Data Set

I thought it would be instructive to see what happens when the data set is a constant. Again, using 10 segments of length 0.1 each and $\lambda = 10$, I fitted mid-segment data of amplitude 1. Figure 4.5 shows that the fitting curve goes through every data point as it should, but has strong ripples at the ends. There is nothing inherent in the data that says how, or even if, the data should go to zero at the ends, but nonzero end values would be unphysical for a wire current. Ripples like this are similar to the Gibbs phenomenon seen when using a Fourier series to fit a square wave, another case of using smooth functions to fit a jump discontinuity.

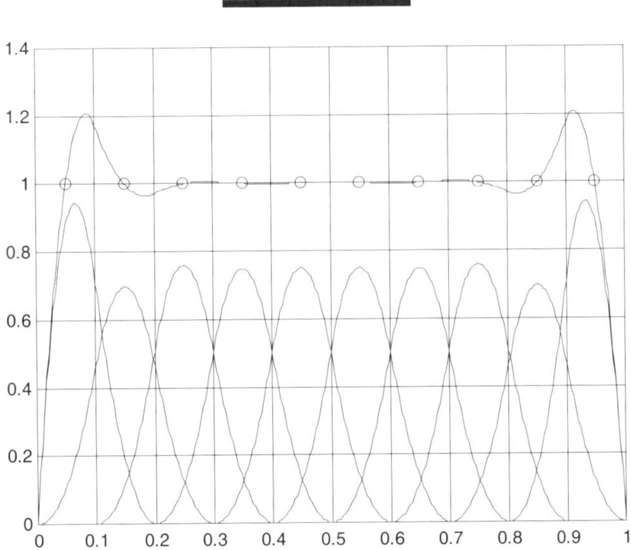

Figure 4.5: Use of NEC2 basis functions to fit a constant-valued data set. The circles mark the values being fitted. The wire length is $\lambda/10$.

Figures 4.4 and 4.5 show the individual basis functions, multiplied by the solution or fitting coefficient, for the two fitting cases. Each of these functions has the form $I_k f_k(s)$, in which I_k is the solution or fitting coefficient and $f_k(s)$ is the basis function, as used earlier. On each segment k of the wire, a main function $m(s_k')$, and some tail functions, $b(s_k')$ or $a(s_k')$, have their domains (are nonzero). Each of these functions has the same three-subcomponent form, so their sum also has this form. For a three-segment piece of wire, the total current on the middle segment is:

$$I(s_k') = A_k + B_k \sin(\beta s_k') + C_k \cos(\beta s_k') \tag{4.36}$$

with

$$A_k = I_{k-1}A_a(s_k') + I_k A_m\,(s_k') + I_{k+1}A_b(s_k')$$
$$B_k = I_{k-1}B_a(s_k') + I_k B_m\,(s_k') + I_{k+1}B_b(s_k')$$
$$C_k = I_{k-1}C_a(s_k') + I_k C_m\,(s_k') + I_{k+1}C_b(s_k')$$

The functional argument notation is used to emphasize that all the A,B,C coefficients are for subfunctions that have domain on the *k*th segment. The subscript *a* coefficients are from a tail of $f_{k-1}(s)$, the subscript *b* functions are from a tail of $f_{k+1}(s)$. The NEC2 code does calculate these overall current coefficients after the solution for the I_k coefficients.

Up to this point, I have been writing as if all problems were single-wire antennas. Of course this is not the case, and the NEC codes have been written to handle junctions of many wires. In the case of a multiwire junction, each basis function will have a main subfunction, and a tail subfunction for each of the other connecting segments. The boundary condition on current is still Kirchoff's law—the sum of the currents into the junction is zero. The charge condition is more complicated. There is an overall charge quantity associated with the junction, and each current derivative is related to this charge quantity by a function of its wire radius. The general problem has been solved both for interior basis functions and for basis functions defined at open wire ends [1,3]. The codes could have been written to set up and numerically solve the equations for each basis function, but the authors pushed analysis to save computation time.

4.2.2 Applied Field Models

NEC offers the user two ways to model an applied field for transmitting. Conceptual models for them are shown in Figure 4.6. In (a), the physical picture is that of a parallel-plate capacitor replacing the driven segment. The capacitance is not specified or included in the computation. The justification for this source model is that the computed electric field near the wire behaves this way, when this model is used on the middle of three equal-length segments. This model gives very stable results for input impedance over a wide range of d/a and d/λ values.

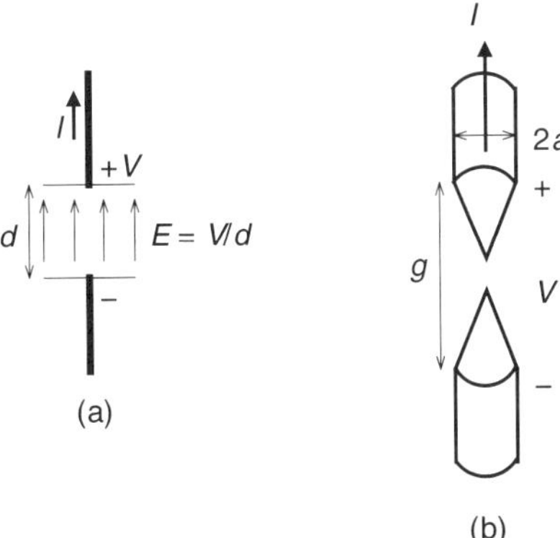

Figure 4.6: Illustrations of NEC source models. (a) Uniform applied field, where *d* is the segment length. (b) Bicone or charge discontinuity model, where *a* is the wire radius and *g* < *d* is a gap length.

The model in (b) is intended for low-frequency and thin-wire applications. It is an effort to model the physical charge discontinuity at a feedpoint. Through various approximations, the authors use the following relation for the change in current slope at the feedpoint:

$$\Delta\left(\frac{dI}{ds}\right) = \frac{-j\beta V}{60\left[lm\left(\frac{d}{a}\right)-1\right]} \tag{4.37}$$

You can see immediately that this is problematic for d/a close to $e = 2.718...$ In NEC, the source is modeled as being applied at the segment end where the current is assumed to enter. This means that a basis-type function with a specified derivative can be used over the source segment and the following segment to model the jump discontinuity. Specifying the beginning slope determines all of the coefficients in the function. Let the local names for the source and following segments be 1 and 2, with lengths d_1 and d_2, and local coordinates s_1' and s_2'. Then the function is:

$$\begin{aligned} f(s) &= \{m(s_1'), a(s_2')\} \\ m(s_1') &= A_m + B_m \sin(\beta s_1') + C_m \cos(\beta s_1') \\ a(s_2') &= A_a + B_a \sin(\beta s_2') + C_a \cos(\beta s_2') \end{aligned} \tag{4.38}$$

For a slope of Q at the left end of segment 1, the coefficients are:

$$\begin{aligned}
R &= (Q/\beta)/\left[\sin(\beta d_1)+\sin(\beta d_2)-\sin(\beta(d_1+d_2))\right] \\
A_a &= R\left(1-\cos(\beta d_1)\right) \\
A_m &= R\left(\cos(\beta(d_1+d_2))-\cos(\beta d_1)\right) \\
B_a &= -A_a \sin(\beta d_2/2) \\
C_a &= -A_a \cos(\beta d_2/2) \\
B_m &= A_m \sin(\beta d_1/2)+(Q/\beta)\cos(\beta d_1/2) \\
C_m &= -A_m \cos(\beta d_1/2)+(Q/\beta)\sin(\beta d_1/2)
\end{aligned} \tag{4.39}$$

Since the normal basis functions have continuous slope at segment junctions, a slope jump is obtained by adding the function (4.38) to the currents. Since this function is known, its fields are also known and added to the applied-field side of the integral equation. The boundary condition enforced by the integral equation is zero tangential electric field at the mid-segment points. The slope-discontinuity function will have fields at all segments, giving a full applied-field (forcing) vector.

Example 4.3 **Fitting a Short-Sine Function with a Slope Discontinuity**

Returning to Figure 4.4 in Example 4.1, you can see that if the fitting curve had reached 1 at $s = 0.5$, the slope on the left would be about 2 and that on the right would be –2. This is a slope jump of –4. I have written MATLAB programs for the derivative-discontinuity basis function, necbasisdd.m, and to use it in data fitting, necfitdd.m. Using these, I fitted the same data as in Example 4.1, plus requiring a slope jump of –4 at $s = 0.5$. The results are shown in Figure 4.7. The negative-valued basis function supplies the slope jump, and the other basis functions have to compensate for its presence to fit the data. This destroys the symmetry of the basis functions seen in Figure 4.4.

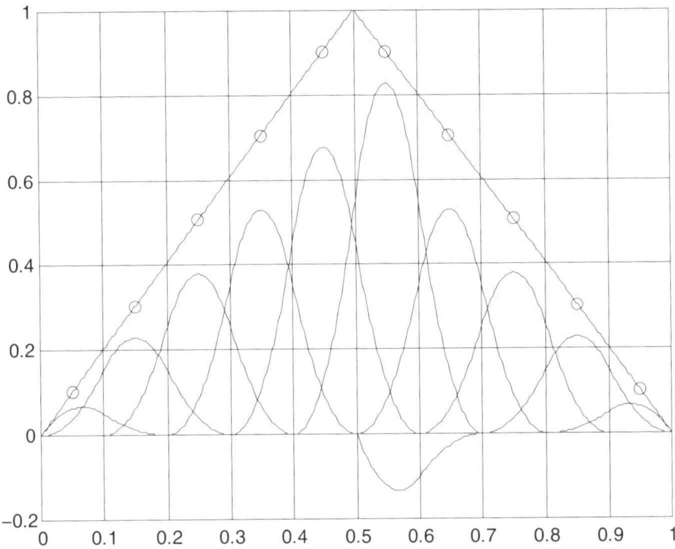

Figure 4.7: NEC basis functions and the fitting curve for the short-sine data set and a slope discontinuity. The circles mark the data points. The slope change requested was −4, at s = 0.5.

4.2.3 Solving the Integral Equation

Now we are ready to discuss solving equation (4.3). You have already seen that NEC represents the current as the weighted sum of basis functions, one basis function per segment in the structure. This means that there are as many unknowns as there are segments. While there are other possibilities, NEC enforces (4.3) at the center of each segment. This method is called *point matching*, and the enforcement points are called the *match points*. This gives the right number of equations for the unknowns. If the uniform-applied-field model is used and the structure has only one applied voltage, then all segment centers will have zero field, except where the source voltage is applied. On the right-hand side of (4.3), the I_k (**r**') is replaced by the unknown coefficients times the basis functions that have domain on segment k. If we represent the unknown coefficients as J_k, $k = 1 \ldots N$, for N segments and basis functions $f_k(s)$, the J_k factors out of the various integrals. When all are evaluated, each J_k is multiplied by a number that depends on the segments for which its basis function has domain, and the segment i for which the field is being enforced. The ith equation has the form

$$\sum_{k=1}^{N} z_{ik} J_k = E_i = E_a(s_i) \tag{4.40}$$

The collection of the N equations forms a matrix equation, which is solved in NEC by the LU (Lower-triangle, Upper-triangle) decomposition method. The matrix of the $[z_{ik}]$ is called variously the impedance matrix, the reaction matrix, or the interaction matrix. Most of the time taken to find the input impedance and radiation pattern for an antenna is spent in calculating the element values and inverting the impedance matrix.

4.3 Using NEC in the Command Window

From the user's point of view, NEC has three main parts:

1. An input file reader that translates a text file into numbers that describe the structure to be simulated;
2. A computational engine that sets up and solves equation (4.3) and then calculates any requested performance data;
3. A program that writes the structure description, certain geometry analysis results, error messages, and the requested performance data to a text file.

NEC was written in FORTRAN 77. It was written for punched-card input and very-wide-sheet line printer output. The modern versions read from and write to text files, but retain the original formatting. Each line of the input file must begin with a two-letter code that identifies the type of data or command to which the following numbers should be applied. In the older versions, the input numbers had to be separated by commas, but now spaces are good enough. Data units in NEC are basic and not user-specified, with one exception. The geometry unit is meters. Electrical units are farads, ohms, henries, siemens, and megahertz (F, Ω, H, S, and MHz). The exception is that a GS, geometry scaling, command can be included to multiply the geometry data by a number before it is interpreted as meters.

The format for describing a wire is:

GW tag segs x1 y1 z1 x2 y2 z2 radius

"tag" is an integer that uniquely identifies the wire. "segs" is the number of segments the wire is divided into, an integer. The *x*, *y*, *z* values are the coordinate of each end of the wire. The reference direction for current and applied sources is from end 1 to end 2. "radius" is the wire radius, also in meters. The floating-point numbers can be in either fixed-point (0.0081) or exponential (8.1e–3) format. Figure 4.8 illustrates some of these ideas. On the left, you see a wire with general coordinates, and the reference direction for current flows from end 1 (x_1,y_1,z_1) to end 2 (x_2,y_2,z_2). The middle wire is vertical. If the reference direction is desired to be from top to bottom and the wire has a 4-mm radius, its description might be:

```
GW 21 1 2.0 3.0 5.0 2.0 3.0 0.0 0.004
```

The right-hand wire is parallel to the *x* axis. If the reference direction is desired to be from +*x* to –*x*, its description might be:

```
GW 276 5 3.0 9.0 0.0 -5.0 9.0 0.0 0.004
```

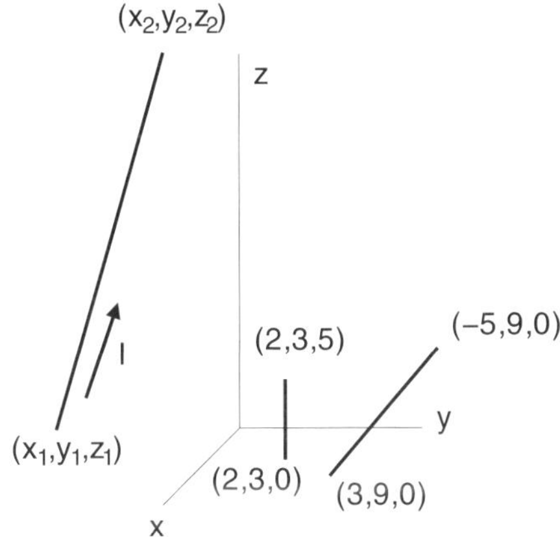

Figure 4.8: Illustration for wire descriptions in NEC.

The order of commands in the input file is not arbitrary. Any descriptive comments come first, with each line started by CM. After the comments, a CE (comment end) line must follow. The CE line may also contain comments. The text in the CM and CE lines will appear in the top of the output file. Then the geometry lines are listed, ending with a GE, geometry end, command. Then program control commands are listed, such as EX for excitation, PT for printing currents (writing them to the output file), RP for radiation pattern, and LD for loads. The last line is EN for end. All the commands and their uses are described in the NEC User's Manual, in the NEC folder on the accompanying CD-ROM.

Example 4.4 Half-Wave Horizontal Dipole in NEC

Suppose we want to study a half-wave dipole in free space operating at 100 MHz. A half wavelength is 1.5 m. As pointed out in Chapter 2, the pattern of an antenna is fixed with respect to the antenna, but its expression depends on the antenna's orientation in a coordinate system. The following wire list represents the antenna as being parallel to the *y* axis, 2 m above the *x*-*y* plane. Only one wire is needed, and it is divided into an odd number of segments so that the driving voltage can be centered in the antenna by placing it in the middle segment.

```
CM Half-wave dipole in free space.
CE
GW 1 9 0 -0.75 2 0 0.75 2 0.0008
GE 0
EX 0 1 5 1 141.4 0 1
FR 0 21 0 0 88 1
PT -1
RP 0 19 2 1000 0 0 10 90
EN
```

The EX command parameters are 0 = uniform-applied-field source, 1 = wire number, 5 = segment on the specified wire to be driven, 1 = a place holder in this case, 141.4 = real part of the peak source voltage (equivalent to 100V rms), 0 = imaginary part of the source voltage, and 1 = normalization factor. The FR command specifies the frequencies the source is to have. The parameters are 0 = arithmetic (rather than multiplicative) progression, 21 = number of frequencies, 0,0, are place holders, 88 = start frequency in megahertz, and 1 = frequency increment in megahertz. This command will cause the input impedance to be listed in a table from 88 to 108 MHz. The PT –1 command suppresses current output. The default is PT –2 which causes output of the current in all segments. This is what happens if no PT command is included. The Radiation Pattern, RP, parameters are 0 = normal mode, 19 = number of values of θ, 2 = number of values of ϕ, 1000 = XNDA to be discussed later, the next two zeros are the start values for θ and ϕ in degrees, 10 = step size in degrees for θ, 90 = step size in degrees for ϕ. This command will cause two elevation sweeps to be done, one at $\phi = 0$ and the other at $\phi = 90°$. NEC will write the gain in dB for vertical and horizontal polarization, the total gain, and some polarization parameters.

The power pattern reported by NEC is not the normalized power function described in Chapter 2, although you can get this information as well. What is reported with the XNDA values given above is the gain as a function of direction. It is the power density in that direction divided by (transmit power/area of sphere). This is expressed as:

$$G(\theta,\phi) = \frac{S(\theta,\phi)}{P_{\text{in}}/4\pi R^2} \tag{4.41}$$

If you set $N = 1$, a table of normalized values will be printed in addition to the unnormalized gains. If you set $D = 1$, unnormalized directive gain is printed instead of the gain of (4.41). Directive gain is defined as in (4.41) except that P_{in} is replaced by $P_{rad} = P_{in} - P_{loss}$. P_{loss} is the power lost in the structure. It does not include ground loss when an earth ground is present.

The frequency response request and the pattern request will combine to produce a lot of output because the pattern data will be generated for each frequency. To tidy up this situation, one can use the XQ (execute) command to separate the frequency sweep for impedance and get the pattern at the mid-band frequency.

Listing 4.1: hwdplfs.nec

```
CM Half-wave horizontal dipole in free space.
CE
GW 1 9 0 -0.75 2 0 0.75 2 0.0008
GE 0
EX 0 1 5 1 141.4 0 1
PT -1
FR 0 21 0 0 88 1
XQ
FR 0 1 0 0 100 0
RP 0 19 2 1000 0 0 10 90
EN
```

After a frequency sweep command, only the RP and XQ commands will cause execution of the commands the program has read to that point.

Depending on the version of Windows you're running, the command window is accessed by running the DOS prompt (command.exe or cmd.exe). It would be useful to put a shortcut icon to this program on your desktop. I assume you have made a folder C:\NEC and that you have copied the NEC2DXS*.EXE files from the CD-ROM to this folder. Now suppose you have written one of the above listings in a text editor. Either the DOS Edit program or Notepad can be started from the command prompt. You may save it with any name as long as the name has no more than eight characters, not including a three-character extension. So suppose you save the listing as IN, and you want the results in a file named OUT. The most direct way to run the program is to type the name of one of the executables and hit Enter. For this case, the structure has only nine segments, so the program with the smallest memory capacity should be used. Type

```
nec2dxs500↵ (↵=Enter key)
```

The program will prompt you for the names of the input and output files. As an investigator and designer, I modify and run and modify and run .... a structure description many times. It's important to minimize the amount of typing or mouse motion needed to get through an iteration in this loop. The first step is to have the .exe program read the in and out file names. This is accomplished by putting the input and output file names in a file. Let's name it **necrun**. All it needs is the two names and two carrier returns.

Listing 4.2: necrun

```
in↵
out↵
```

Now, at the command prompt, you could type

```
nec2dxs500 <necrun↵
```

Of course, after the program runs, you will want to examine the results and maybe modify the input file. You could write a batch file called **gonec.bat** as follows:

Listing 4.3: gonec.bat

```
echo off
notepad in
nec2dxs500 <necrun
notepad out
```

When exercising any of the listings in this part of the chapter, it is best to copy the listing file to the file **in,** and then make modifications on **in** rather than the original. For example, type

```
copy hwdplfs.nec in↵
```

after you have written listing 4.1. Notepad has the advantage over Edit that you can copy lines out of Notepad and paste them into other Windows applications. However, in Windows 98 Notepad does not have enough capacity for a NEC output file so Edit must be used. When you are all done, you can use a Windows-based word processor to get the data you need out of the NEC output file. Now you can type gonec↵ to start a pass; this opens **in**, you can modify it, save it, and press Alt-F4 to close Notepad. The program will run automatically and then open Notepad with **out** so you can see the results. Use Alt-F4 to close Notepad again, and the batch file terminates. You can start another pass by pressing F3, which echoes the previous typing, and then ↵.

Example 4.5 **Adding a Ground Plane**

In NEC the *x*-*y* plane is also the ground plane when not modeling in free space. NEC4 allows for wires below ground, but NEC2 does not. Since our horizontal dipole is at two meters height, there is no problem. A ground plane is requested by changing the GE command parameter from 0 and adding a GN command to describe the type of ground and material properties. The following listing places the dipole over "average" ground, with a conductivity of 0.005 S/m and a relative permittivity of 13. The RP command has also been changed to sweep θ from 0 to 90°. Even without the change, the program will not calculate the gain for elevations below ground.

Listing 4.4: hwdpl.nec

```
CM Half-wave horizontal dipole over ground.
CE
GW 1 9 0 -0.75 2 0 0.75 2 0.0008
GE 1
GN 2 0 0 0 13 0.005
EX 0 1 5 1 141.4 0 1
PT -1
FR 0 21 0 0 88 1
XQ
FR 0 1 0 0 100 0
RP 0 10 2 1000 0 0 10 90
EN
```

Example 4.6 **Monopole on a Perfectly Conducting Plane**

This listing changes the half-wave dipole to a quarter-wave monopole over an infinite PEC plane. This is done by changing the ground description to GN 1, and rewriting the GW command. Also, the source is applied at segment 2 to simulate driving the monopole at its base.

Listing 4.5: qwpec.nec

```
CM Quarter-wave vertical on a PEC plane.
CE
GW 1 9 0 0 0 0 0 0.75 0.0008
GE 1
GN 1
EX 0 1 2 1 141.4 0 1
PT -1
FR 0 21 0 0 88 1
XQ
FR 0 1 0 0 100 0
RP 0 10 2 1000 0 0 10 90
EN
```

When you run this listing, you will see that the two elevation sweeps are the same, so the third parameter in the RP command could have been changed from 2 to 1.

NEC has provisions to model lumped and distributed loads through the LD command. The loads are placed in series with the segments to which they are applied. If a source and several loads are applied to the same segment, they are all treated as being in series.

Example 4.7 Series-Tuning the Monopole

The impedance of the monopole at 100 MHz has a reactance component of $j25.53\Omega$. This can be series-resonated with a capacitance of 6.23 pF. LD 0 specifies a series RLC load. Following the load type indicator, the next three integers follow a pattern common in NEC. The first specifies a wire tag number, and the next two the start and finish segments for the loading. In our case, we want to load wire 1 and only the source segment, 2, so the entries are 1 2 2. If no element is wanted for either R or L, we put 0 in those positions. The result is Listing 4.6.

Listing 4.6: Tuned quarter-wave monopole.

```
CM Quarter-wave vertical on a PEC plane.
CM Tuned at 100 MHz.
CE
GW 1 9 0 0 0 0 0 0.75 0.0008
GE 1
GN 1
LD 0 1 2 2 0 0 62.3e-12
EX 0 1 2 1 141.4 0 1
PT -1
FR 0 21 0 0 88 1
XQ
FR 0 1 0 0 100 0
RP 0 10 1 1000 0 0 10 90
EN
```

Next, we can model the effect of finite wire conductivity by using LD 5. Assume a conductivity of 26 MS/m for aluminum. The following version of the command will apply this loading to all wires.

```
LD 5 0 0 0 26e6
```

Before adding this to the listing, make careful note of the maximum gain, efficiency, and impedance reported by running the listing above. The loss addition will make small changes in each of these.

Loading can be applied to a single segment, to a range of segments on a wire, or to selected wires. This is useful if you have a structure made of different metals. The selection is done by using the tag and segment-range integers in the LD command.

4.4 Modeling Guidelines

There are a number of approximations in NEC, and there are numerical hazards for modeling small structures, as pointed out in earlier sections. Some guidelines are given in the NEC manual, and some others have been published by users (e.g., [5,6]). These are collected in Table 4.3.

Table 4.3: Dimensional guidelines. Generally, violation of any of these restrictions will be flagged as an error in model-checking programs. The references give more restrictive conditions whose violation will produce a warning.

Parameter	*Relation*	*Reasons*
1. Open wire length, L and wire radius a	$L > 20a$	Neglect of end caps.
2. Wire radius a	$a < \lambda/20\pi$	Thin-wire approx.
3. Max. segment length, d	$d < \lambda/2\pi$	Basis function limitation.
Min. d	$d > 10^{-4}\lambda$	Single-precision arithmetic.
	$d > 10^{-8}\lambda$	Double-precision arithmetic.
4. a vs. d, NEC2 NEC4	$a < d/2$ $a < 2d$	Same as 2.
5. Axial separation of parallel wires, r	$r > 3a$	Same as 2.
Wire Junctions		
6. Angle θ of two intersecting segments, $d_1 < d_2, \theta < 90°$	$a_2 < (d_1/2)\sin\theta - a_1\cos\theta$ $\sin\theta > 2(a_1 + a_2)/d_1, \theta$ small	Match point of one wire should be outside the volume of the other wire.
7. Max/min segment length	$d_{max}/d_{min} < 5$	Charge distribution accuracy.
8. Max/min wire radii	$a_{max}/a_{min} < 10$	As above.
Wire near ground in NEC2		
9. Distance from wire center to ground ($z = 0$), h	$h > 3a$, $(h^2 + a^2) > 10^{-12}\lambda$	

Wire mesh model of surface		
Square grid, s by s		
10. Grid length, s	$s < 0.1\lambda$	Field leakage.
11. Mesh wire radius, a	$a = s/(2\pi)$	Wire area equal to surface area.
12. General cell shape perimeter length, L	$0.04 < L/\lambda < 1$	Prevent nonphysical circulating currents.

Other cautions are:

- Intersecting wires MUST do so at segment ends. Intersections at wire ends only are recommended.
- The EX 0 source is best applied to the middle of three segments of equal length and radius, and lying on a straight line.
- The EX 5 source is best applied to the junction of two segments of equal length and radius, and lying on a straight line.
- If either source is applied at the base of a monopole over a ground plane, the segment should be vertical, and the connected radial segments should be the same length as the source segment.
- Avoid large radius changes between connected wires. Some versions of the code have a correction algorithm for this condition.
- The number of wires joining at a junction is limited to 30 by the original NEC2 code.
- Although the angle condition in Table 4.3 item 6 prevents match-point intrusion into the next wire space, it really doesn't put enough space between the wires. If $a_x = \max\{a_1,a_2\}$ a condition such as $\theta \geq 2(a_1 + a_2 + a_x)/d_1$ will insure a bit more space.
- At multiple-wire junctions, the segment lengths should be equal, if possible.
- At multiple-wire junctions, the wire radii should be the same, if possible.
- When wires are parallel and close, their segment ends should line up so that the match points line up across the wires.
- Small, voltage-driven loop models fail if the circumference is less than $3 \times 10^{-4}\lambda$ (double-precision, NEC2). Close coupling between wires and small loops should be avoided in NEC.

All of the previous are guidelines, not guarantees of a successful model. There are several internal checks described in the NEC manuals. The easiest one to do is to compute the average gain. The XNDA parameters in the RP command can be used to request the average power gain based on the angle parameters specified in the command. For a lossless structure in free space or over a PEC ground, the value of average gain over the entire space should be 1. The *A* part of XNDA calls for this average if it's 1 or 2. $A = 1$ displays the gain at all the points called for by the angle parameters, and $A = 2$ suppresses this pointwise printing and reports only the average gain. If you set up the integration to use a reasonably fine spacing, say 2 degrees, you will get more points than you want to see, so $A = 2$ is the better choice.

If structure loss is present, the efficiency report should equal the average gain times 100%. If a finitely conducting ground is present, the average gain will be less than the efficiency/100. This is because the power radiated into the ground is not counted in the average gain integral, and it's not part of the structure loss. From equation (4.41), in NEC the average power gain is:

$$G_{avg} = \frac{\frac{1}{\text{Area}}\int_A S(\theta,\phi)\,dA}{\frac{1}{4\pi R^2}P_{in}} = \frac{4\pi R^2}{\text{Area}}\cdot\frac{P_{rad}}{P_{in}} = \frac{4\pi R^2}{\text{Area}}\cdot\frac{R_{rad}}{R_{in}} \tag{4.42}$$

where *A* is the surface, at distance *R*, over which the radiated power density is being averaged. In free space, Area = $4\pi R^2$, and over a ground plane, Area = $2\pi R^2$. If G_{avg} is not efficiency/100 in a free-space test, (4.42) can be used to estimate the correct radiation resistance. It has been suggested [7] that criteria for the quality of a model based on G_{avg} are:

- deviation of less than 5% implies an excellent model.
- deviation of less than 10% implies a usable model.
- deviation less than 20% implies a usable model that needs improving.
- deviation greater than 20% implies an unusable model.

Using G_{avg} to correct the radiation resistance is less reliable for larger deviations.

A second test for the accuracy of a NEC model is to increase the number of segments per wire. The minimum number in the guidelines above is 7 segments per wavelength, but modelers typically start around 20 and go up. In ordcr to get a physically accurate model of an electrically small structure, one is usually using more segments per wavelength anyway. You will find that the resonant frequency, maximum gain, and input impedance (either at a fixed frequency or at resonance) change with the number of segments. In some cases, as the segments are increased by some fixed increment, the changes decrease in magnitude, seeming to settle to fixed values to 4 or 5 figures. This is called *convergence*. Convergence in this sense doesn't always happen. Sometimes a range of segment numbers gives a minimum change, and then increasing the segment numbers causes the change to increase. The best segmentation is the range for minimum change. A particular danger for small structures is that the segment length to wire radius ratio may become too small. For structures with different parts, such as a vertical antenna with top and bottom radials, the segmentation density can be different in the different parts. You have a great deal of flexibility to improve your model, and to get in trouble.

Example 4.8 **Average Power Gain**

Listing 4.7: Tuned quarter-wave monopole, average gain.

```
CM Quarter-wave vertical on a PEC plane.
CM Tuned at 100 MHz.
CE
GW 1 9 0 0 0 0 0 0.75 0.0008
GE 1
GN 1
LD 0 1 2 2 0 0 62.3e-12
EX 0 1 2 1 141.4 0 1
PT -1
FR 0 21 0 0 88 1
XQ
FR 0 1 0 0 100 0
RP 0 10 1 1000 0 0 10 90
RP 0 46 46 1002 0 0 2 2
EN
```

Notice that there is no need for an XQ command between the two single-frequency RP commands. You should run this listing for several values of segment number, and record the gain, input impedance and average power gain. Part of the impedance variation is due to the fact that, as the number of segments is increased, the applied voltage gets closer to the base of the antenna, which should be a current maximum.

4.5 NEC in a Graphical User Interface (GUI)

The advantages of a GUI are the ability to show information in drawings, and the ability to copy and paste information to other Windows applications, especially to reports and articles. There are a number of commercial GUIs available, but we have included a free one on the accompanying CD-ROM called 4nec2, written by Arie Voors [2,4].

NEC is still the computational heart of the GUI application, so the application has to take input in some form, convert it to a NEC input file, run the computation, and then extract the desired data from the NEC output file. In 4nec2 you have the option of using a regular text editor, a built-in text editor with help for the command fields, and a Geometry Editor which gives you a picture of the structure and helps you to build a model with a minimum knowledge of NEC commands. The outputs available are the basic NEC output file, and any number of plots that depend on what data was requested. The Calculate panel allows the user to either run the input file as is, or override its data requests. The computer's function keys can be used to move quickly between editing the input, running the model, and viewing the outputs, saving your wrist. The installation and use are described in files on the CD-ROM, so I will not cover these topics here in any detail. The package is a zip file, and should be unzipped with a Windows-based, rather than a DOS, program so as not to truncate some of the file names.

Even if the input is a *.nec text file, 4nec2 will read and process it before passing it on to the NEC engine. This is because Arie has added two commands to improve the flexibility of use and interface with the built-in optimization programs and loop-evaluation. These are the SY (symbol) command, and use of the single quote as a comment mark anywhere in the file. The symbol command is used to define a variable name with an initial value, which is then used in the NEC commands instead of numbers. One SY command can include multiple assignments, which must be separated by commas. Arithmetic and functional operators can also be used. The functions available are "sin", "cos", "tan", "atn", "sqr", "exp", "log", "abs", "sgn", "fix" and "int". In addition to the usual four arithmetic operators you can use "^", the power operator.

Example 4.9 **80-m Band Dipole**

The 80-m amateur radio band runs from 3.5 to 4 MHz. The following listing describes a horizontal dipole made of #14 wire suspended over average ground. I used 4nec2 to find the value of hlen that minimized the input reactance at 3.75 MHz. Then I used the frequency sweep feature to override the FR command and generate the impedance plots shown in Figure 4.9.

Listing 4.8: dpl80.nec

```
CM "Half-wave" horizontal dipole over ground.
CE
SY hlen=19.175,hgh=2 'Half-length and height above ground.
SY eps=13,sig=0.005
'Average ground rel. perm. and conductivity.
GW 1 9 0 -hlen hgh 0 hlen hgh 0.0008
'Antenna is parallel to y axis.
GE 1
GN 2 0 0 0 eps sig
FR 0 1 0 0 3.75
EX 0 1 5 5 141.42 0
RP 0 1 1 1000 70 0 0 0
'Gain at 20 degrees elevation, in x-z plane.
EN
```

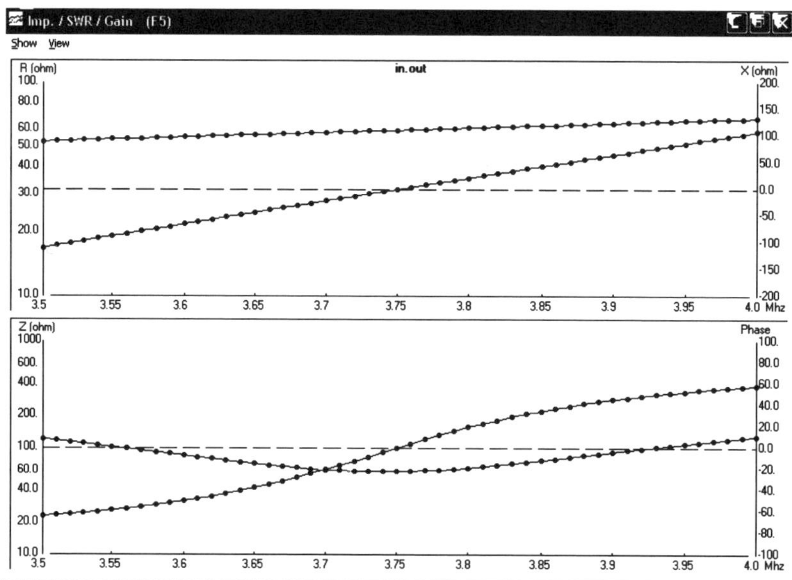

Figure 4.9: Plots of input impedance for Listing 4.8. Alt-PftScr was used to copy the plots.

4.6 Examples from Chapters 2 and 3

We studied the short dipole and the small loop in Chapter 2 and mentioned end-loading the short monopole in Chapter 3. In this section, I give NEC wire lists for these antennas and compare the results with the analytical formulas' results.

4.6.1 The Short Dipole

In Example 2.2, we calculated the input impedance, including the radiation resistance and loss resistance, for a 1-m aluminum tube dipole at 30 MHz. The results are:

$$R_{\text{rad}} = 1.974\Omega,\ X_{\text{in}} = -1{,}307\Omega$$

$$R_{\text{loss}} = 0.01887\Omega$$

$$R_{\text{in}} = 1.993\Omega.$$

$$\text{Efficiency} = 99.05\%$$

$$\text{Directivity} = 1.5 = 1.76\text{ dB}$$

The following listing is for this dipole on the z axis. It is the final version after doing the calculation for no loss and with loss, and segment numbers from 9 through 45. Note that the driven segment, which must be at the dipole center, is nsrc = (segs + 1)/2 and that segs must always be odd. The first RP command gets the maximum gain for this orientation, which is also the directivity in the absence of loss. The results are in Table 4.4.

Listing 4.9: Short dipole in free space.

```
CM Short dipole in free space.
CE
GW 1 45 0 0 0 0 0 1 0.006
GE 0
EX 0 1 23 23 100 0
LD 5 0 0 0 26e6
PT -1
FR 0 1 0 0 30 0
RP 0 1 1 1000 90 0 0 0
RP 0 46 46 1002 0 0 2 2
EN
```

Table 4.4: NEC2 calculations for a 1-m long, 12-mm diameter aluminum dipole in free space at 30 MHz, λ = 10 m.

No loss.

segs	Z_{in}	G_{max}	*APG*	*Eff, %*
9	2.14497E + 00 – j1.34766E + 03	1.78	0.9958	
15	1.99120E + 00 – j1.29171E + 03	1.77	0.9985	
25	1.87503E + 00 – j1.24930E + 03	1.77	0.99948	
35	1.80816E + 00 – j1.22505E + 03	1.78	0.99974	
45	1.75950E + 00 – j1.20747E + 03	1.78	0.99984	

With loss, 26 MS/m.

segs	Z_{in}	G_{max}	*APG*	*Eff, %*
9	2.16501E + 00 – j1.34764E + 03	1.72	0.98667	99.07
15	2.00983E + 00 – j1.29170E + 03	1.73	0.98928	99.07
25	1.89256E + 00 – j1.24928E + 03	1.73	0.99023	99.07
35	1.82506E + 00 – j1.22504E + 03	1.73	0.99049	99.07
45	1.77594E + 00 – j1.20745E + 03	1.74	0.9906	99.08

The impedance doesn't converge, but it keeps on getting smaller as the number of segments is increased. In effect, the impedance is accurate to one digit. The error in the no-loss gain is about 1%. The efficiency is good to 0.02%. Notice that the APG with loss converges to the correct efficiency value. These results illustrate the difficulty of simulating a highly reactive antenna. The input impedance depends on one number—the current in the driven segment—so it is most sensitive to inaccuracy. The gain and efficiency are the results of integration over the structure and so are less sensitive to individual errors in the current. The final example in this section will show how much better a near-resonant model behaves.

4.6.2 Small Loop in Free Space

Next, we revisit the loop of Example 2.3. From that and following examples we have, again at 30 MHz,

$$R_{rad} = 1.92278\Omega,\ L = 2.829\ \mu H,\ C_{loop} = 4.09\ pF.$$

$$R_{loss} = 0.1778\Omega,\ \text{Efficiency} = 91.64\%$$

$$Z_{in} = 6.06 + j906\Omega.$$

The following listing models the loop using the GA (geometry arc) command. It generates an arc in the x-z plane, centered on the y axis. The syntax is:

GA tag nw R α_s α_f r

It generates a series of nw straight lines inscribed inside the arc. The arc's radius is R and the wire radius is r. α_s is the starting angle of the arc, measured up from the x axis toward the z axis (a left-handed direction when thinking of the xyz coordinates as a right-handed system); α_f is the finishing angle. "tag" is the number of the first wire. The runs start with 16 wires because the actual circumference of the polygon is within 1% of the modeled circle.

Listing 4.10: Small loop in free space.

```
CM Small loop in free space.
CE
GA 1 16 0.5 0 360 0.006
GE 0
EX 0 1 1 1 100 0
PT -1
FR 0 1 0 0 30 0
RP 0 1 1 1000 90 0 0 0
RP 0 46 46 1002 0 0 2 2
EN
```

Table 4.5: 1-m diameter loop of 12-mm diameter aluminum tubing at 30 MHz, in free space. *No loss.*

Wires	R_{in}	X_{in}	G_{MAX}, *dB*	*APG*
16	6.21	820.5	1.51	1.0346
32	7.23	872	1.55	1.0224
48	7.63	893	1.55	1.0177
64	7.87	905.8	1.56	1.0153
80	8.043	915	1.56	1.0138

Adding loss to the model yields, with 80 wires, APG = 0.952, efficiency = 93.9%, G_{MAX} = 1.28 dB, and $Z_{in} = 8.57 + j915.5\Omega$. Note that the ratio of the no-loss R_{in} to the with-loss R_{in} is 0.936, the same as the efficiency/100. As with the dipole, there is no sign of real convergence, and the resistance values are high. Also like the dipole, the reactance moves to more inductance.

4.6.3 End-Loaded Short Dipole

For our final example, we test the idea of making a small dipole look like a capacitor, as described at the end of Chapter 3. I chose to look for a design that would resonate between 3 and 4 MHz, ($75 < \lambda < 100$ m) and be 4 m high. I used four 16-m long radials at each end of the vertical radiator. In the following listing, the first wire is the radiator, and the next eight wires are the two sets of radials.

Listing 4.11: End-loaded vertical in free space.

```
CM End-loaded vertical in free space.
CE
GW 1 51 0 0 0 0 0 4 0.0004
GW 2 204 0 0 0 16 0 0 0.0004
GW 3 204 0 0 0 0 16 0 0.0004
GW 4 204 0 0 0 -16 0 0 0.0004
GW 5 204 0 0 0 0 -16 0 0.0004
GW 6 204 0 0 4 16 0 4 0.0004
GW 7 204 0 0 4 0 16 4 0.0004
GW 8 204 0 0 4 -16 0 4 0.0004
GW 9 204 0 0 4 0 -16 4 0.0004
GE 0
EX 0 1 26 1 316 0 1
FR 0 8 0 0 3 0.1
PT -1
RP 0 46 46 1002 0 0 2 2
EN
```

The radials were made four times longer than the vertical to make it easy to specify their segments to have the same length as the segments on the vertical. It turned out that the resonant frequency is near 3.3 MHz. In Table 4.6, there are two impedance entries, one for 3.3 and the other for 3.4 MHz. This is to show the stability of the data with segment density.

Table 4.6: End-loaded dipole in free space, center-fed. For each row, the entries in the *R* and *X* columns are for 3.3 and 3.4 MHz. The APG figures were the same, rounded to the extent shown.

Vert. segs.	*R, Ω*	*X, Ω*	*APG*
5	1.585 1.69	–1.187 14.96	1.007
11	1.59 1.696	–0.742 15.36	1.004
21	1.592 1.699	–0.386 15.69	1.002
51	1.594 1.7015	0.0883 16.12	1.0013

The radiation resistance for a 4-m long uniform-current radiator is 1.53Ω. Again, the computed value is higher, but only by 4%. I believe this example illustrates that NEC does much better with an electrically short structure if it's near resonance.

References

[1] Burke, G. J. and A. J. Poggio, *Numerical Electromagnetics Code (NEC)-Method of Moments*, NOSC TD 116, January 1981. My copy was obtained from the National Technical Information Service (U.S.A.).

[2] http://www.si-list.org/swindex2.html, http://www.robomod.net/mailman/listinfo/nec-list
(Use a search engine to find the current home pages for NEC2 and NEC-LIST if these don't work.)

[3] Burke, G. J., *Numerical Electromagnetic Code—NEC-4, Method of Moments*, Lawrence-Livermore National Laboratory, 1992.

[4] Voors, Arie, 4nec2@gmx.com. Contact Arie with questions, but download his software from the NEC2 archive location given in [2].

[5] C. W. Trueman and S. J. Kubina, "Verifying Wire-Grid Model Integrity with Program CHECK," ACES Journal, vol. 5, no. 2, pp. 17–42, Winter, 1990.

[6] Cebik, L. B., "Some Basic Guideline Graphics for NEC," #58 in a series of articles to be found at www.cebik.com and www.antennex.com.

[7] Cebik, L. B., "The Average Gain Test," #20 in a series of articles to be found at www.cebik.com and www.antennex.com.

Chapter 4 Problems

Section 4.2.1

4.1 Derive equations (4.4.18–26), for the basis subfunction coefficients.

4.2 Solve equations (4.8–15) using the NEC2 condition (4.16).

4.3 Derive the small-argument versions for the subfunction coefficients, (4.27–30).

4.4 Derive the subfunction coefficients for a two-segment basis function. Use (4.17) as the amplitude-setting condition. The main subfunction is next to an open wire end, so the current must be zero, but the charge is unspecified.

4.5 Repeat Example 4.1, but use nine segments. Comment on the results and differences.

4.6 Repeat Example 4.2, but with 20 segments. Comment on the results.

Section 4.2.2

4.7 Derive the slope-discontinuity equations (4.39).

4.8 Find approximations for (4.39) when $d_1 = d_2 = d << \lambda$.

4.9 Define an offset-triangle function on $0 < z < 10$. The triangle shall be zero at $z = 0$ and 10, and 1 at $z = 3$. Use 10 segments, $\lambda = 10$, and generate a data set at the segment midpoints. Find the slope jump at $z = 3$. Use necfitdd.m (or your equivalent in a programming language or another math environment) to fit the triangle. Plot the weighted basis functions and the fitting curve. Use at least 101 points.

Section 4.2.3

4.10 In equation (4.40), if the units of J_k are amperes, what are the units of z_{ik}?

Section 4.3

4.11 Type and run listings 4.1 and 4.4. Describe the differences. Note especially the input impedance and radiated power at 100 MHz. Explain the effect of the ground on these numbers.

4.12 Write and run Listing 4.5 for the 1/4-wave monopole. Compare this to the results for Listing 4.1. Cite the relevant theory for the changes in impedance and gain.

4.13 Rewrite Listing 4.5 to have two wires, one short one for the source, and one long one for the remaining antenna. Rewrite the listing so the only data out is the impedance at 98 MHz. Study the impedance as the source wire is made shorter, both with one segment and with three segments for the source wire. In the latter case, the source should be in the middle segment.

4.14 Run Listing 4.6, the tuned monopole, with and without loss. Relate the changes in gain, input resistance, and efficiency to each other.

4.15 Derive the condition for intersecting wires in item 6 of Table 4.3.

Section 4.4

4.16 Modify Listing 4.7 so that the source drives the segment at the base of the monopole, change the number of segments to 10, and delete the first RP command. Run the listing and retune it for 100 MHz, if necessary. Run the program and record the approximate resonant frequency, input impedance, and average power gain for increments of 10 segments, until either convergence seems likely, or the segment length/wire radius limit is reached. Comment.

Section 4.5

4.17 Modify Listing 4.8, the 80-m dipole, to include variable segmenting. Be sure to set the source at the middle segment. Use the Evaluate feature to find the input impedance as segments are incremented by 10. Does convergence occur? If so, what is the minimum number of segments for impedance to be stable to three figures?

4.18 From Figure 4.9, what are the resistance, reactance, magnitude and phase at 4 MHz of the antenna input impedance?

Section 4.6.3

4.19 Convert Listing 4.11, the end-loaded short dipole, for use with 4nec2's SY command. Make the antenna height, radial length, wire radius, and wire segments variables. Minimize the number of variables needed to change the antenna's description. Add copper loss to the model.

(a) Use 25 segments for the vertical, and test your model at 3.3 MHz.

(b) Set the wire radius to AWG #14. Retune the antenna to 3.5 MHz by varying the radial length.

(c) Make a table of efficiency and input impedance at 3.5 MHz as the wire radius is changed from #18 to 1/4" tube, in steps of 2 in the gauge. Comment on the trends.

CHAPTER 5

Programmed Modeling

5.0 Introduction

As is often said, numerical experiment and design require a lot of trips around the modify-run-examine loop. In the last chapter, I showed how the SY command in 4nec2 can be used to simplify the modify step. In the next section, you will see that there are NEC commands that generate a whole set of wire descriptions. After that, I describe using a programming language, C++, to generate a NEC input file from a relatively simple file that describes the geometry, operating conditions, and desired output.

5.1 Using Wire-List Generators in NEC

Simple antenna structures can be readily described and altered in a direct NEC-command text file. A more complex structure, such as a spiral antenna on a box, would take a large number of GW commands, and altering the geometry would require altering a great many GW commands. The writers of NEC recognized this problem and provided some commands to make this job easier. You have already seen the GA (Geometry Arc) command in the last chapter. The first I describe here is the GM (Geometry Move) command. This command generates a specified number of duplicates of a previously described structure that are rotated and translated incrementally from each other and the original. Suppose you want a wire list to describe a short monopole on a PEC ground, with four radials at the top. You have to have a GW line to specify the vertical wire and another GW line to specify the first radial. Then you can use a GM command to duplicate the first radial. The listing might look like this.

```
CM Short HF monopole.
CE
GW 1 11 0 0 0 0 0 4 0.0008
GW 2 44 0 0 4 16 0 4 0.0008
GM 1 3 0 0 90 0 0 0 2
GE 1
GN 1
EX 0 1 2 2 100 0
FR 0 26 0 0 3.5 0.02
PT -1
XQ
EN
```

The syntax of the GM command is as follows:

```
GM taginc copies RoX RoY RoZ dX dY dZ tagstart
```

If "copies" is zero, no duplicates are generated, but the original structure is moved. In the present case, we want three copies. Each Ro- number specifies a rotation about the named axis, in degrees. If the structure in question does not have a point on an axis for which a nonzero rotation is specified, it will still be swung around that axis. An advantage of a GUI is that you can see a drawing of the result of the instruction and decide if what happened is what you intended. In the present case, we want to duplicate the specified radial at 90° intervals around the *z* axis. The dX, dY, dZ specify the translational increment. The order of execution of these motions is in the order listed, from left to right. Each copy is moved from its predecessor by the same amounts. The structure duplicated is from tagstart to the GM command. Within NEC, the copies are assigned the tag numbers of the original structure plus taginc times the copy number. You can see this in the output file. In the present case, we are copying wire 2, and there's no danger of making duplicate tag numbers, so we only really need to make taginc = 1. When you read the output file, you'll see that the tag numbers for the radial copies are 3, 4, and 5.

If you want to vary the structure in any way, you have to change at least two numbers in the listing. In 4nec2, you can use the SY command to minimize the number of items you have to change to accomplish a geometry variation.

Listing 5.1: HF monopole on a PEC.

```
CM Short HF monopole.
CE
SY hgh=2, len=16, rpt=3, rad=0.0008
SY sgv=11, sgr=int(0.5+len*sgv/hgh)
'Approx. equal segment lengths on vertical and radial.
SY ang=360/(rpt+1)
SY f1=3.5, f2=4, df=0.1, nof=1+int(0.5+(f2-f1)/df)
'f1=start freq., f2=stop freq., df=freq. inc.
'df must not be zero.
GW 1 sgv 0 0 0 0 0 hgh rad
GW 2 sgr 0 0 hgh len 0 hgh rad
GM 1 rpt 0 0 ang 0 0 0 2
GE 1
GN 1
EX 0 1 2 2 100 0
FR 0 nof 0 0 f1 df
PT -1
XQ
EN
```

The GR (Geometry Rotate) command is a variation on the GM command. It is simpler, more restricted, and more powerful within its restrictions. It takes a previously defined structure and rotates copies of it to fill a cylindrical space. The rotation is around the *z* axis, so the base structure can't have segments lying along or across the *z* axis because this will cause multiple segments to have illegal intersections or occupy the same space. When a completely rotational structure is defined this way, symmetry flags are set to enable filling and factoring a much smaller impedance matrix. Listing 5.2 generates a structure that could have been made with two GA commands, but if so done the symmetry would not have been used.

The GR command only uses two integers; the tag increment is the first and the number of copies of the base structure, including the original, is the second. In Listing 5.2, the base structure is the first two wires of a polygon approximation for each loop.

The symmetry flags are unset if the GR command is followed by a GW command, if lumped-element loading is used, or if the GR command is followed by a GM command, no harm in itself, and then a ground description which makes the situation physically unsymmetrical.

Listing 5.2: Coaxial loops.

```
CM Coaxial loops in free space.
CM Demo GR command.
CE
GW 1 1 1 0 0 0.7071 0.7071 0 0.001
GW 2 2 2 0 0 1.4142 1.4142 0 0.001
GR 2 8
GE 0
EX 0 1 1 1 100 0
EX 0 2 1 1 100 0
FR 0 1 0 0 10.0 0
XQ
EN
```

In Listing 5.2, each wire starts out on the *x* axis and goes out to a point in the first quadrant at the same radius and 45° out from the *x* axis. The GR command then generates seven more copies of this chord across the arc.

The other command that NEC uses to generate symmetric structures is GX (Geometry refleXion). This command causes a base structure to be reflected along each coordinate axis. The syntax is:

```
GX inc xyz
```

"inc" is the tag increment, as with the GR command. "xyz" is three digits, either 1 or 0, grouped together without spaces. Each position that has a "1" causes reflection along that axis, starting from the *z* axis and working to the left. The things that will unset the symmetry flags are the same as with the GR command, with two exceptions. If successive GX commands are used, only the last one can set symmetry flags. Lumped loads can be used if their positions have the same symmetry as the last GX command.

Listing 5.3: HF-band rhombic antenna.

```
CM HF-band rhombic antenna.
CE
GW 1 50 -346 0 75 0 200 75 0.0008
GX 1 100
GX 2 010
GS 0 0 0.3048
GE 1
GN 2 0 0 0 13.0 0.005
LD 0 2 1 1 650.0
```

```
LD 0 4 1 1 650.0
EX 0 1 1 1 100 0
EX 0 3 1 1 -100 0
FR 0 1 0 0 7.1 0
PT -1
RP 0 1 2 1000 70 0 0 180
EN
```

Figure 5.1 shows the sequence of operations implied by Listing 5.3. The process begins with a wire running from a $-x$ position to a $+y$ position. The first GX command reverses the signs of the x coordinates of each segment in the wire, so its reflection starts on the $+x$ axis and goes to the same point on the $+y$ axis. The next GX command takes all the y values in this structure and reverses their signs to produce the final result. The shape of the final wire structure could have been produced by GX 1 110, but this would have set symmetry flags for both the x and y axes, which will be unset when the resistors are applied symmetrically only with respect to the y direction.

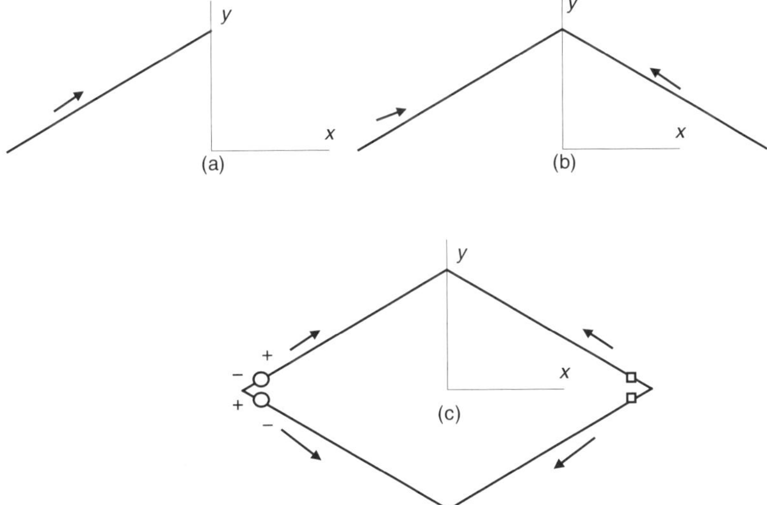

Figure 5.1: Illustration of the effect of the GX command. The arrows show the reference direction for current on each wire. (a) Base wire. (b) GX 1 100, wire reflected along *x* axis. (c) GX 2 010, wires in (b) reflected along *y* axis. In (c), the top two wires have tag numbers 1 and 2, and the bottom two wires have tag numbers 3 and 4, left to right in both cases. The circles represent the locations of the applied voltages, and the squares represent the locations of the terminating resistors. The two voltage sources represent a balanced feed; the lower source is opposite the reference direction so it has to have a minus sign in the EX command in the listing.

The numbers in the GW command are feet. The GS command is used to convert them to meters by multiplying them by 0.3048 m/ft. This antenna would be both physically and electrically large. It is a type that has a wide bandwidth and high directivity. It's a form of a class called travelling-wave antennas, in which a wave is launched by the source, travels along the wires losing power by radiation, and the remaining power is absorbed by the terminating resistors. Ideally, there is no reflected wave along the wires, and this property contributes to its wide bandwidth. I designed this model by starting with a side length of 400' and an angle of 30° off the x axis. I wrote a version using SY commands and used 4nec2 to find the terminating resistances that would minimize the antenna input reactance. When you run this listing, you will find that the model has a high gain, high front-to-back ratio, and high input resistance. NEC reports the impedance seen by each source in the model, so the actual impedance is the sum of the two. If built, the antenna would be fed through a quarter-wavelength of "ladder-line," an open parallel-wire transmission line with insulating spacers at regular intervals. Commercial versions of this line have a wave impedance of 450Ω.

The GR and GX commands are rather specialized in their possible applications. When they can be used to set symmetry flags, a great deal of time can be saved in an optimization process. Otherwise, the GM command is very flexible and can save a lot of writing, especially when combined with SY commands in the 4nec2 environment.

5.2 Using Code to Generate a Wire List

To begin, let's look at the monopole on a PEC ground plane. What we want is a program that reads an input file for a few numbers that describe the geometry and writes a NEC input file as its output. Listing 5.4 is an example of such a program. You might say to yourself, "Look at all that overhead code!" but this is an example, not something I would do in practice for such a simple structure. I also chose not to use the GM command, which would have simplified the program, because I want to illustrate the fact that using code allows you to generate a variable number of wire statements. Using code also allows you to test for issues in the geometry whether or not they are tested in the NEC program itself. In this listing, I have a test on the radial spacing to keep down the wire overlap. The program is arranged to let the user specify the start, stop, and step frequencies so that the number of frequencies doesn't have to be calculated by the user. A test on the step frequency is included both to prevent a divide-by-zero error, and to give an easy way to switch from a frequency sweep to a single-frequency run.

Listing 5.4: The monopole in C++.

```
#include <stdio.h>
#include <stdlib.h>
#include <float.h>
#include <math.h>
#include <iostream.h>
#include <iomanip.h>
#include <fstream.h>

void main()
{
ifstream inp("mono.in",ios::nocreate);
ofstream outp("mono.nec",ios::ate);
outp<<setprecision(6);
float h,len,vsglen,rsglen,ang,rad;
int k,nrad,vsg,rsg,nf;
float f1,f2,df;
//Read monopole height, no. of segments, wire radius.
inp>>h>>vsg>>rad;
//Read no. of radials and length.
inp>>nrad>>len;
//Read start, stop, and step frequencies.
inp>>f1>>f2>>df;
ang=2*M_PI/nrad;//Angle between radials.
vsglen=h/vsg;//Vertical segment length.
rsg=floor(0.5+len/vsglen);//No. of radial segments.
rsglen=len/rsg;//Radial segment length, approx. same as vsglen.
//Test spacing between radials.
if (ang<6*rad/rsglen)
   {
   cout<<
   "Wire radius too large for no. segs and radial length.\n";
   return;
   }
//Write output file.
outp<<"CM Monopole on PEC Ground PLane.\n";
outp<<"CE\n";
//Write vertical wire.
outp<<"GW 1 "<<vsg<<" 0 0 0 0 0 "<<h<<" "<<rad<<endl;
//Write radials.
for (k=0;k<nrad;k++)
```

```
    {
    outp<<"GW "<<(k+2)<<" "<<rsg<<" 0 0 "<<h
        <<" "<<(len*cos(k*ang))<<" "<<(len*sin(k*ang))<<" "<<h
        <<" "<<rad<<endl;
    }
outp<<"GE 1\nGN 1\n";//Specify ground plane.
outp<<"EX 0 1 1 1 316 0 1\n";
if (df>0)
    {
    nf=1+floor(0.5+(f2-f1)/df);
    }
else
    {
    nf=1;
    }
outp<<"FR 0 "<<nf<<" 0 0 "<<f1<<" "<<df<<"\n";
outp<<"PT -1 \n";
outp<<"RP 0 46 46 1002 0 0 2 2 \n";
outp<<"EN \n";
inp.close();outp.close();
}
```

A sample for the file **mono.in** is given in Listing 5.5. Observe that both the source code file and the data input file can be read to find out how to use the program, but the information in the .in file is more compactly given and is where you need it when you're going to change the numbers.

Listing 5.5: Input file for wire-list generator mono.exe.

```
2 10 0.0008
4 10
3.5 4 0.1

This file, mono.in, is the input file for mono.exe which gen-
erates a NEC input file mono.nec. All numbers are in NEC
standard units. By row, they are:

1. Monopole height, no. segments in vertical, wire radius.
2. No. of radials, length of radials.
3. Start, stop, and step frequencies. If the step frequency is
   zero, only the start frequency will be used.
```

Many antenna systems are arrangements of common shapes so it is useful to have code blocks that generate vectors of coordinate values that can be used in a wire-list generator. In C++ this is best done by using functions. The inputs to these functions are the basic geometry parameters of the shapes, and the outputs are passed by reference to the coordinate vectors. Listing 5.6 shows a function used to build a NEC description of a disc represented by a wire grid. This function is one of several collected in the file **meshes.h**. All of these functions use objects in a vector class, defined in the file **dbarrays.hpp**.

Listing 5.6: The disc() function.

```
void disc(int Nr, int Na, double R, vector& x1, vector& x2,
   vector& y1, vector& y2, vector& r)
{
double th,dth,cth,sth,r1,dR,rw,Rc,L,chord;
vector rs(Nr),Rs(Nr+1),A(Nr);
int j,k,inx,koff;
k=2*Nr*Na;x1.resize(k);x2.resize(k);
y1.resize(k);y2.resize(k);r.resize(k);
dth=2*M_PI/Na;Rs(0)=0;chord=2*sin(0.5*dth);
r1=R/Nr;rs(0)=r1;Rc=r1;dR=r1;
sth=0.25*sin(dth)/M_PI;A(0)=r1*r1*sth;
for (k=1;k<Nr;k++)
   {
   rs(k)=dR;
   Rs(k)=Rc;r1=Rc;Rc+=dR;
   A(k)=(Rc*Rc-r1*r1)*sth;
   }
Rs(Nr)=Rc;
//Radials.
for (k=0;k<Na;k++)
   {
   th=k*dth;cth=cos(th);sth=sin(th);
   for (j=0;j<Nr;j++)
      {
      inx=j+k*Nr;
      x1(inx)=Rs(j)*cth;x2(inx)=Rs(j+1)*cth;
      y1(inx)=Rs(j)*sth;y2(inx)=Rs(j+1)*sth;
      r(inx)=A(j)/rs(j);
      }
   }
//Circles.
koff=Nr*Na;
for (j=0;j<Nr;j++)
   {
   Rc=Rs(j+1);L=chord*Rc;th=0;
```

```
        rw=A(j)/L;
        for (k=0;k<Na;k++)
            {
            inx=koff+k+j*Na;
            x1(inx)=Rc*cos(th);y1(inx)=Rc*sin(th);
            th+=dth;
            x2(inx)=Rc*cos(th);y2(inx)=Rc*sin(th);
            r(inx)=rw;
            }
        }
    double eps=1e-6;
    zerovec(x1,eps);zerovec(x2,eps);
    zerovec(y1,eps);zerovec(y2,eps);
    zerovec(r,eps);
    }
```

The intent of this function is to produce a wire-mesh imitation of a thin disc of radius *R*, centered in the *x*-*y* plane. The user has some control over the quality of the simulation by specifying the number of radial steps, *Nr*, and the number of angular steps, *Na*. These three items are the function inputs. The function outputs are the *x*-*y* coordinates of the wire ends, four vectors, and a vector to hold the corresponding wire radius values. The wire radius is adjusted so that the wire surface area approximates the portion of the disc area the wire is simulating. The area the wire is simulating is a function of its direction, because it is simulating a current density flowing in that direction, multiplied by the disc area covered by that current density. This is illustrated in Figure 5.2.

Figure 5.2: Illustration of an element in the wire-mesh approximation to a disc.

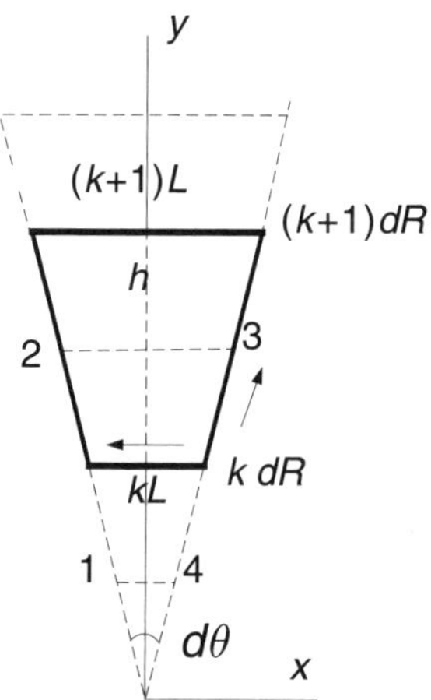

The arrows show the reference direction for current in the angular and radial directions. The illustration shows a section of the grid spanning the *y* axis. The horizontal wire at *kdR* needs to simulate the area enclosed by the figure whose corners are labeled 1-2-3-4. The radial wire running from *kdR* to $(k + 1)dR$ has to simulate the area enclosed by the four-wire cell. *dR* is the radial step size, *R*/*Nr*. $d\theta$ is the angular step size, $2\pi/Na$. *L* is the chord length for the innermost ring,

$$L = 2dR\sin(0.5d\theta) \tag{5.1}$$

The distance between wires in successive rings, *h*, is

$$h = dR\cos(0.5d\theta) \tag{5.2}$$

The area enclosed by 1-2-3-4 is *kLh*. The surface area of the chord wire is $kL2\pi r_k$. This leads to

$$r_k = \frac{h}{2\pi} = \frac{dR}{2\pi}\cos(0.5d\theta) \tag{5.3}$$

as the wire radius for the chord wire in ring *k*. The area enclosed by the mesh cell is $(k + 0.5)Lh$ and the surface area of the radial wire is $dR2\pi r_k$. This leads to a wire radius of:

$$r_k = \frac{(k+0.5)Lh}{2\pi dR} = \frac{k+0.5}{2\pi}dR\sin(d\theta) \tag{5.4}$$

You can see that the chord wires have constant radius because the spacing between them is constant, while the radius wires have to be thicker as they go out from the center because the space between them grows.

After the declarations, the basic geometry calculations are done and then the wire list construction is done in two major nested for-loops. In the //radials block, the outer loop steps through the angles, and for each angle, the inner loop constructs the radial wires. The cosine and sine parts of the expressions are constant inside the inner loop, so calculating them outside this loop provides a little time savings, but the wire radius is increased on each pass through the inner loop. In the //Circles code block, the outer loop sets the radius and the inner loop constructs the wire that makes the chord across each angle increment. This coding is typical of the functions that construct wire-lists for the various shapes. The shapes are broken down into their simpler geometric parts, and a code block is written for each of these simple parts. Care must be taken to get the indexing into the output vectors right, so a calculation has been done for the total number of wires in each subshape. At the end of the function, another function, zerovec(), is used to replace small numbers with zeros. This step makes the wire list easier to read and tidier.

The final listing in this chapter is a program that reads an input file, calls a number of shape-generating functions, and writes the NEC input file for an antenna on a box. The antenna has a radiating element that may either be straight or coiled, and is topped by a thick plate (or thin cylinder, depending on your point of view). A particular instance is shown in Figures 5.3 and 5.4 and described by the input file vlcoil.in below.

Listing 5.7: vlcoil.in

```
3 1
5 21
0.4 1200 1800 0 57e6
3 20 1
12.0 4 121
0 20 1
8.0 4 103
1
2 6 8
-7.5 7.5 -7.5 -7.5
-22.5 -22.5 22.5 -22.5
10 10 10 100

This, vlcoil.in, is the data file for vlcoil.exe which gen-
erates the NEC wire list for a circular vlcoil over a
rectangular box. Dimensions are in mm. The box is described
by the coordinates of four corners, three at the bottom, and
the top corner above the inner of the three. The box must
have a wire end on the z axis to connect to the antenna.

The items are, by row:

1. No. of bodies, 2 for no box, 3 with a box. No. turns in
   the coil.
2. Coil radius, coil height.
3. Antenna wire radius. Start, stop , and step freq., MHz.
   Conductivity. If the step freq.=0 only the start freq.
   is used. If conductivity=0, perfect conductors are used.
4. No. radial steps, no. angle steps, no. of height
   steps,for the top plate.
```

```
5. Top plate radius, height, z coordinate of bottom
   surface. 6-7 are the same as 4-5, for the bottom plate.
   If no. radial steps=0, no bottom plate is included.
8. Average Power Gain flag.
   1=run APG.
   0=gain at 0 elevation and 0 and 90 azimuths.
   For the box:
9. No. steps in x,y,z directions.
10. x corner coordinates.
11. y corner coordinates.
12. z corner coordinates.
```

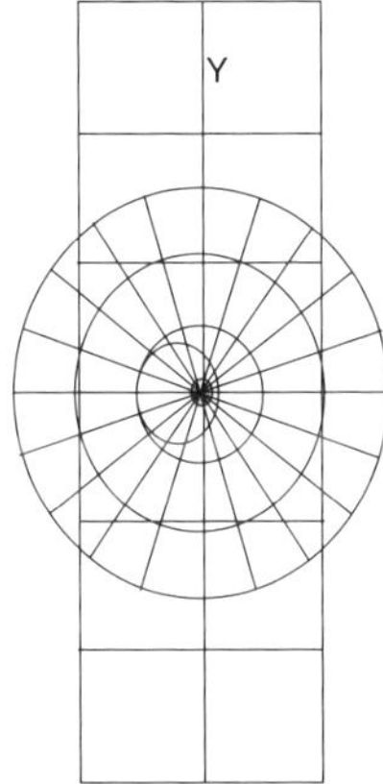

Figure 5.3: Top view of wire model for volume-loaded coil on a box.

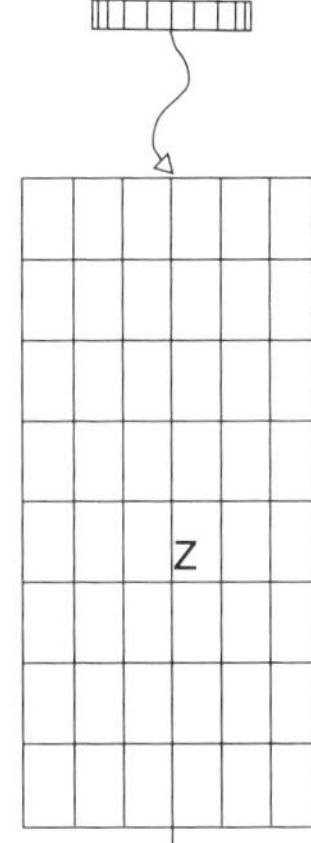

Figure 5.4: Broadside view of wire model for volume-loaded coil on a box.

Listing 5.8: vlc.cpp

```
/*Program for generating a NEC wire
list for a cylindrical VLSD. The parameters are
read from file vlcoil.in
Written by D. B. Miron, June 2000.
*/
#include <stdlib.h>
#include <float.h>
#include <math.h>
#include <iostream.h>
#include <iomanip.h>
#include <fstream.h>
```

```
#include "dbarrays.hpp"
#include "meshes.h"

void main()
{
ifstream inp("vlcoil.in",ios::nocreate);
ofstream outp("vlc.nec",ios::out|ios::trunc);
outp<<setprecision(6);
vector dat(27);
int k,nf,nx,koff,APG;
int nbxs;
double Rc,Hc,Nt,rwc;
int nr1,na1,nz1,nr2,na2,nz2,nbx1,nbx2,nbx3;
//Coil vectors.
vector xt1,xt2,yt1,yt2,zt1,zt2;
//Upper plate vectors.
vector xdu1,xdu2,ydu1,ydu2,zdu1,zdu2,ru;
//Bottom plate vectors.
vector xdb1,xdb2,ydb1,ydb2,zdb1,zdb2,rb;
//Box corners.
vector xcbx(4),ycbx(4),zcbx(4);
//Box vectors.
vector x1bx,x2bx,y1bx,y2bx,z1bx,z2bx;
double Ru,Rb,Hu,Hb,zu,zb,rbx,rw,f1,f2,df,sig;
inp>>nbxs>>Nt; //No. bodies, no. coil turns.
inp>>Rc>>Hc; //Coil radius and height.
inp>>rwc>>f1>>f2>>df>>sig;//Wire radius,
   //start, stop, and step freqs., conductivity.
rw=coil(Rc,Hc,Nt,rwc,xt1,xt2,yt1,yt2,zt1,zt2);
if (rw!=rwc) printf("Coil wire radius changed to %.2f\n",rw);
inp>>nr1>>na1>>nz1;//No. steps in radial, angular and z
   //directions for top plate.
inp>>Ru>>Hu>>zu;//Top plate radius, height, and
   //z coordinate of its bottom.
inp>>nr2>>na2>>nz2;//Steps for bottom plate.
inp>>Rb>>Hb>>zb;//Dims. and z for bottom plate.
if (nr2==0) {zb=zu-Hc;Hb=0;}//If no bottom plate.
if (Hc!=zu-zb+Hb) zu=zb+Hb+Hc;//Correct input data
   //in favor of coil height.
//Make top plate.
cylinder(nr1,na1,nz1,Ru,Hu,xdu1,xdu2,ydu1,ydu2,zdu1,zdu2,ru);
```

```
//Make bottom plate, if present.
if (nr2>0) cylinder(nr2,na2,nz2,Rb,Hb,xdb1,xdb2,ydb1,ydb2,zdb1,
zdb2,rb);
inp>>APG;
//Make box, if present.
if (nbxs>2)
   {
   for (k=0;k<15;k++) inp>>dat(k);
   nbx1=dat(0);nbx2=dat(1);nbx3=dat(2);
   for (k=3;k<7;k++)
      {
      xcbx(k-3)=dat(k);
      ycbx(k-3)=dat(k+4);
      zcbx(k-3)=dat(k+8);
      }
   rbx=box(nbx1,nbx2,nbx3,xcbx,ycbx,zcbx,x1bx,x2bx,y1bx,
      y2bx,z1bx,z2bx);
   }
outp<<"CM VLcoil, "<<Nt<<" turns. \n";
if (nbxs>2)outp<<"CM Cylindrical plate over a box with wire
between,\n";
outp<<"CM in free space, base-fed.\n";
outp<<"CE \n";
koff=xt1.getsize();
double Hoff=zb+Hb+0.5*Hc;//z offset for coil wires.
//The coil z vector is centered at z=0.
for (k=0;k<koff;k++)
   {
   outp<<"GW "<<(k+1)<<" 1 "<<xt1(k)<<" "<<yt1(k)<<" "
      <<(zt1(k)+Hoff)<<" "<<xt2(k)<<" "<<yt2(k)<<" "
       <<(zt2(k)+Hoff)<<" "<<rw<<endl;
   }
koff++;nx=xdu1.getsize();
if ((nbxs>2)&&(nr2>0))//Put a wire between the box and the
      //antenna.
   {
   outp<<"GW "<<koff<<" 1 0 0 "<<zcbx(3)<<" 0 0 "<<zb
      <<" "<<sqrt(rb(0)*rbx)<<"\n";koff++;
   }
for (k=0;k<nx;k++)
   {
```

```
    outp<<"GW "<<(k+koff)<<" 1 "<<xdu1(k)<<" "<<ydu1(k)<<" "
        <<(zu+zdu1(k))<<" "<<xdu2(k)<<" "<<ydu2(k)<<" "
         <<(zu+zdu2(k))<<" "<<ru(k)<<endl;
    }
koff+=nx;
if (nr2>0)
    {
    nx=xdb1.getsize();
    for (k=0;k<nx;k++)
        {
        outp<<"GW "<<(k+koff)<<" 1 "<<xdb1(k)<<" "<<ydb1(k)
            <<" "<<(zb+zdb1(k))<<" "<<xdb2(k)<<" "<<ydb2(k)<<" "
            <<(zb+zdb2(k))<<" "<<rb(k)<<endl;
        }
    koff+=nx;
    }
if (nbxs>2)
    {
    for (k=0;k<x1bx.getsize();k++)
        {
        outp<<"GW "<<(k+koff)<<" 2 "<<x1bx(k)<<" "<<y1bx(k)
            <<" "<<z1bx(k)<<" "<<x2bx(k)<<" "<<y2bx(k)<<" "
            <<z2bx(k)<<" "<<rbx<<endl;
        }
    }
outp<<"GS 0 0 0.001 \n";//meters/mm.
outp<<"GE 0\n";//GN 1\n";
outp<<"EX 0 3 1 1 141.42135 0 1\n";
if (sig>0) outp<<"LD 5 0 0 0 "<<sig<<" 1 \n";
if (df>0)
    {
    nf=1+floor(0.5+(f2-f1)/df);
    }
else
    {
    nf=1;
    }
outp<<"PT -1 \n";
if (APG)
    {
    outp<<"FR 0 "<<nf<<" 0 0 "<<f1<<" "<<df<<"\n";
```

```
    outp<<"RP 0 91 181 1002 0 0 2 2 \n";
    }
  else
    {
    outp<<"FR 0 "<<nf<<" 0 0 "<<f1<<" "<<df<<"\n";
    outp<<"RP 0 1 2 1000 90 0 2 90 \n";
    }
  outp<<"EN \n";
  inp.close();outp.close();
  }
```

I am an engineer first, a user of mathematics second, and a numerical programmer third. While I try to use "good programming practice" when I find it, I don't claim to write supremely efficient or readable code. I learned C++ in 1998 in order to write a special-purpose modeling program, and I'm still learning it. With these caveats in mind, if you have a strong programming background I encourage you to regard my work as a starting point and go on from there.

In Listing 5.8, after the declarations and other overhead code, the input file is read sufficiently far to allow the coil() function to be called. This function is written to allow considerable flexibility. If the number of turns is 0, a straight-wire model is generated. If the number of turns is negative, a left-handed coil is generated, otherwise the coil is right-handed. The coil starts on the $-z$ axis at $-Hc/2$, winds out to full radius in a half turn, winds up to make the specified number of turns minus one half, and winds back in at the top in another half turn to $z = +Hc/2$. The number of turns is a decimal to allow fine adjustment of the coil. It can also be less than 1. The requested wire radius is tested to see if it's too large, and reduced if it is.

The structure modeled can consist of two plates with a coil between, one plate and coil on a box, or two plates with coil between over a box. In the last case a wire is run from the center of the bottom plate down to the box. So the next bit of code reads the input file for the plate modeling and physical parameters. The plates are simulated as wires modeling surfaces, using the equal-area rules given in the discussion of disc() above. In fact, disc() is called twice by the cylinder() function to form the top and bottom surfaces of the plate. So the user specifies the number of wires in each coordinate direction for the cylinder and the corresponding wire radii are calculated and returned by the function. In Figure 5.3, you can see that this instance is for three radial steps that generates three approximate rings, and 20 angle steps around the disc. In Figure 5.4, you can see that only one height (z direction) step is used. For the bottom plate, the number of radial steps is used as a flag to say whether it is present or not.

The first parameter in vlcoil.in is a leftover from previous codes as far as its values are concerned. "nbxs" stands for "number of boxes" and in the present program it specifies that the antenna is on a box if it's greater than 2. The box in this case is 15 mm by 45 mm by 90 mm high, the approximate size of a cordless phone radio box I bought in 1996. The function box() was written to place a box anywhere in space, using the coordinates of four corners to specify it. In this case, the box is centered on the z axis so the antenna will sit in the center of the box top. To get a connection at the center of the box top, there has to be a wire end there, so the numbers of steps in the x and y directions have to be even integers.

After all the wire vectors have been generated, the output file is written. Each part of the structure is written in a separate for{ } loop. The radiating wire is written first because I want to be able to specify the excitation segment without calculating where it is in the wire list. A parameter that's read before the box parameters is the APG flag. The Average Power Gain test is quite useful in deciding the density of wires in different parts of the model.

The above discussion is about a particular example of writing code to generate a wire-list model of an antenna structure. I made some choices in the program sequence that are good programming practice, and some choices that reflect the way I like to work. Chief among the latter is that I generated all the structure vectors before I started output to the .nec file. This requires the definition and storage of more variables than if I had done a generate-and-write sequence for each part of the structure, but I feel more comfortable doing it the way I did. Again, it's important to keep track of the counting for the wire numbers in the GW lines, and I felt it was easier to do that by writing the output file straight through. Finally, to the impatient reader, I discuss the motivation for and performance of this antenna in the next chapter.

Chapter 5 Problems

Section 5.1

5.1 Write Listing 5.1 as a NEC input file. Use the 4nec2 Optimize function to tune (minimize Xin) the antenna at 3.9 MHz using the radial length for 4, 8, and 16 radials.

5.2 Write a .nec file similar to Listing 5.2 for two parallel loops of equal radius and equal wire radius. Use SY commands to make the number of copies, and hence the chord end two coordinates, variable. Use a variable for the space between the loops. For a 1-m loop radius, 1-mm wire radius, and a frequency of 10 MHz, find the spacing that minimizes the reactance for 36-chord loops, using the Optimize function of 4nec2. Record the time. Next, write another .nec file to describe the same structure using GA commands, run the optimization process from the same starting values, and record the time again.

5.3 Write a version of Listing 5.3 using symbols for the leg length, angle and terminating resistance. Use the Optimize function in 4nec2 to maximize F/B and minimize Xin simultaneously with equal weighting. Use the angle and resistance as the variables to be adjusted.

Section 5.2

5.4 Rewrite the function disc() so that the side lengths in a cell of the mesh are more nearly equal. This can be done by making dR a variable, or making the number of angle steps different in each ring. The second strategy is more difficult because each successive ring will have more radius lines whose inward ends will terminate on circles, so the circle wire will have to be split.

5.5 Write a function called plate() which returns wire vectors for a plate centered in the *x,y* plane. The user is allowed to specify the plate dimension and the number of steps in one direction. The function must adjust the number of steps in the other direction to make the mesh cells as square as possible. The function must return a wire radius vector.

5.6 Write a program to generate the .nec file for a straight-wire antenna on the corner of a radio box. The program must read a file that specifies the wire height, radius, number of segments, the box dimensions, and degree of segmentation. Frequencies and pattern request must also be read from the .in file.

CHAPTER 6

Open-Ended Antennas

6.0 Introduction

This chapter presents some ideas and performance results for antennas that are variations on the monopole. Some will contain coils, and so will look like loop variations, but they have an open end at which the current must be zero. Loop variations are presented in the next chapter. The things that can be done to improve a monopole performance for a given space and operating wavelength are:

1. Make it thicker.
2. Increase its capacitance by top loading.
   (a) Disc or radial wires.
   (b) Volume loading.
3. Make it resonant
   (a) by adding an inductor at the base or higher up,
   (b) by making it with enough wire wound in the available space to make it $\lambda/4$,
   (c) by making the radiating element a coil and top-loading it with enough C.

Method 3(b) is wave resonance, whereas 3(a) is circuit resonance by adding a nonradiating inductor, and 3(c) is circuit resonance by making the radiating element also an inductor. Examples of each method are examined with respect to efficiency, impedance matching, and bandwidth. All designs for monopoles on a PEC plane are scalable with wavelength, but antennas on a radio box are a different story. The radio box is part of the antenna, but its dimensions are usually not determined by this fact; rather, they are made as small as will accommodate the electronics. Sometimes the box is a significant fraction of a wavelength, and sometimes it isn't.

Most of the antennas discussed in this chapter conform, more or less, to a cylinder. If we imagine a parallel-plate capacitor with a dipole antenna wire between the plates, the plates being circular with radius a and separation h_d, we can approximate the structure's circuit impedance as follows. Assuming a uniform current along the wire gives us the radiation resistance, from (2.39), $R_r = 80\pi^2(h_d/\lambda)^2$. The capacitance can be approximated with the low-frequency uniform-charge-density formula $C = \epsilon_0 Area/h_d = \epsilon_0 \pi a^2/h_d$. Wheeler [1, ch. 6] defined the radiation power factor for a small cylindrical capacitive antenna as $p_e = k_e R_r/X_c = k_e R_r \omega C = k_e/Q$. Here k_e is a fudge factor described as being slightly larger than 1. Setting the fudge factor to 1, and using the expressions for R_r and C just given, one arrives at Wheeler's expression for the radiation power factor (inverse element Q) of a small antenna occupying a cylindrical volume,

$$p = \frac{1}{6\pi} \cdot \frac{Area\, h_d}{(\lambda/2\pi)^3} = \frac{1}{6} a^2 h_d \beta^3 \tag{6.1}$$

This implies a quality factor

$$Q_W = \frac{6}{a^2 h_d \beta^3} \tag{6.2}$$

The radius of a sphere just fitting around our cylinder is $r^2 = a^2 + (h_d/2)^2$. Using this in Chu's minimum Q expression (3.17), we have

$$Q_C = \frac{8}{\beta^3\left(4a^2 + h_d^2\right)^{3/2}} + \frac{2}{\beta\left(4a^2 + h_d^2\right)^{1/2}} = \frac{8}{(\beta h_d)^3\left(1 + \frac{4a^2}{h_d^2}\right)^{3/2}} + \frac{2}{\beta h_d\left(1 + \frac{4a^2}{h_d^2}\right)^{1/2}} \tag{6.3}$$

It is clear that these two expressions for Q are not equivalent. Wheeler's Q is a statement about what is possible for a single-reactance-type small antenna, while Chu's Q is a minimum based on the theoretical possibility of radiating a single spherical mode. I will use them as benchmarks for comparison with the various examples in the following sections. In particular, five cases are shown in Table 6.1 for future use.

Table 6.1: Q values for cylindrical antennas in free space. $h_d = 0.1\lambda$.

a/λ	Q_C	Q_W
0	35.4	---
0.0125	32.5	1548
0.025	25.9	387
0.05	13.7	96.75
0.1	4.3	24.2

A monopole on a PEC ground plane has half the impedance of a dipole twice its length. Since both the reactance and resistance are halved, the ratio is the same, and so is the Q. So, for a monopole of height h_m, the Q is the same as for an equivalent dipole of height $h_d = 2h_m$. This changes (6.2) and (6.3) to:

$$Q_W = \frac{3}{a^2 h_m \beta^3} \tag{6.4}$$

$$Q_C = \frac{1}{\beta^3\left(a^2+h_m^2\right)^{3/2}} + \frac{1}{\beta\left(a^2+h_m^2\right)^{1/2}} = \frac{1}{\left(\beta h_m\right)^3\left(1+\frac{a^2}{h_m^2}\right)^{3/2}} + \frac{1}{\beta h_m\left(1+\frac{a^2}{h_m^2}\right)^{1/2}} \tag{6.5}$$

The entries in Table 6.1 also apply for a monopole of height $h_m = 0.05\lambda$.

6.1 Thick Monopoles

I only know of two experimental studies of the monopole that cover a wide range of radius and height values. The first was published in 1945 by Brown and Woodward [1, pp. 4–6:4–9] and in the 1960s by students of R. W. P. King [2]. Figure 6.1 illustrates two different ways to drive a thick monopole. Figure 6.1(a) is a coaxial cable below the ground plane with the center conductor extended to make the antenna. This method was used in both sets of measurements. Figure 6.1(b) shows a relatively narrow coax with its center conductor passing through to the base of a thick cylindrical antenna. This method was also used by Brown and Woodward, although the details of the feed geometry are not given.

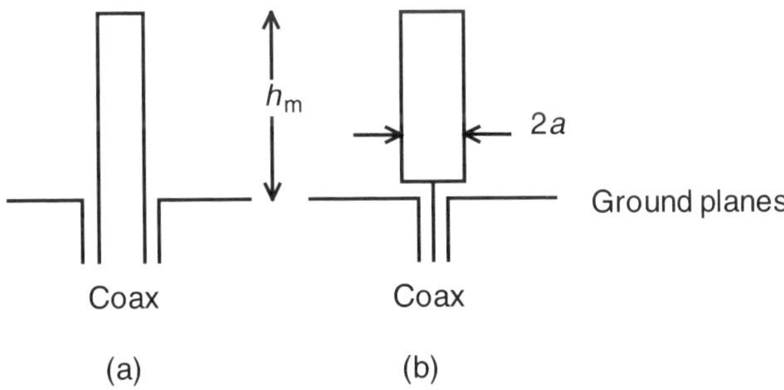

Figure 6.1: Two monopole feed arrangements.

The data in [1] is given as curves in two different formats. The resolution for $h_m < 0.1\lambda$ for the resistance curves is not very good, but one can draw some general conclusions. For the straight-through coax feed system of Figure 6.1(a), the tighter the coax is, the smaller the reactance magnitude is. The feed system of Figure 6.1(b) has a much lower reactance magnitude than the straight-through cases, and also appears to have lower resistance. The data in [2] is given both in curves and tables, and I use it to verify the modeling results of the next section.

6.1.1 Modeling Thick Monopoles

NEC is based on thin-wire theory, so a model of a thick cylinder has to be built from multiple wires, just like modeling a surface. The general cylinder() function mentioned in Chapter 5 allows for currents in vertical, radial, and angular directions, and I used this in a preliminary study of the monopole fed as in Figure 6.1(b). However, from symmetry it is apparent that there should be no angular currents. Also, a way must be found to model the extended coax monopole of Figure 6.1(a). King [2] observes that the capacitance of the open cable end is in series with the antenna. The tighter the fit, the smaller this excess reactance should be. I modeled this case by using vertical wires with multiple segments for the cylinder side, and put a voltage generator in the bottom segment of each wire. For N wires, the total antenna current is N times the current reported for each source, and the net impedance is $1/N$ of that reported for each source. For convenient reference, I call this the edge-driven monopole (EDM). King and his students measured the admittance of monopoles with open tops, flat tops, and hemispherically capped tops, but I only modeled the flat-top case with a modified version of disc(), symdisc(), to take out the angular currents. Symdisc() generates radial wires, with each radial wire made up of short wires of equal length and increasing wire radius to approximate the equal-area rule for modeling a surface. The overall symmetrical cylinder function is called symcylinder(). The model of the drive system in Figure 6.1(b) I call the center-driven monopole (CDM). Figure 6.2 illustrates a model for the EDM.

An objective of the segmentation is to make the side cells, as if there were angular-directed wires, as square as possible, and the radial wire segments equal to the vertical wire segment length. If N_a is the number of angle steps (vertical wires), the angle step is $d\theta = 2\pi/N_a$, the arc-length step on the cylinder side is $a \times d\theta = 2\pi a/N_a$, the number of vertical segments should be $N_v \approx h_m/(2\pi a/N_a) = (h_m/a)(N_a/2\pi)$. Likewise, the number of radial wire segments is $N_r \approx N_a/2\pi$. The expressions are approximate because the various Ns must be integers. Note that $N_v/N_r \approx h_m/a$.

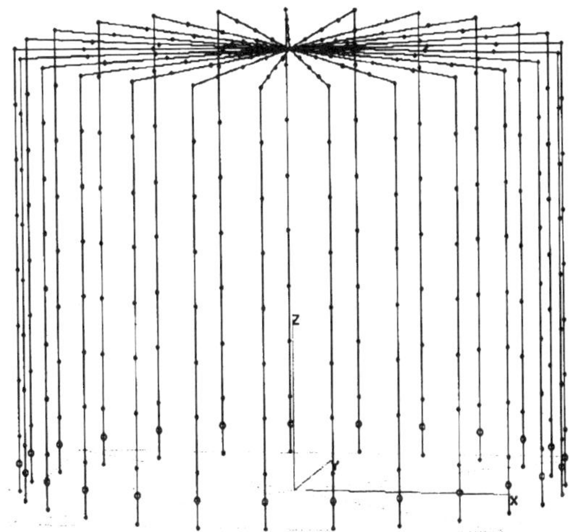

Figure 6.2: Thin-wire model for a thick monopole. Overall, it is 50 mm high by 50 mm diameter, driven at 300 MHz so that λ = 1 m. It is an EDM, edge-driven as the limiting case for the extended thick coax center conductor. The large dot at the bottom of each vertical wire is a voltage source. The small dots are at segment ends.

As in previous examples, the impedance changes with segment density. The following table shows the progression for an edge-fed monopole similar to one of King's cases. The average power gains for all the computations reported in this section are within 2±0.01.

Table 6.2: Impedance as a function of segment density for an edge-fed monopole. h_m = 0.1λ, a = 0.0509λ. $\Delta Z = 100[Z_{k+1}-Z_k]/Z_k$.

N_r	N_a	N_v	Z, Ω	ΔZ, %
3	20	6	4.72–*j*33.7	–
4	25	8	4.16–*j*31.7	-6.17–*j*0.82
5	32	10	3.78–*j*30.3	-4.54–*j*0.59
6	38	12	3.51–*j*29.2	-3.59–*j*0.45
7	45	14	3.3 –*j*28.4	-2.94–*j*0.36
8	51	16	3.14–*j*27.7	-2.49–*j*0.29
9	58	18	3 –*j*27.1	-2.15–*j*0.25

The last entry was chosen as the density to use in the following runs, because 60 seems to be the maximum value of the number of wires at a junction in the version of nec2dxs*.exe I'm using.

The next table compares computed values for the edge-driven monopole against King's experimental values. King remarked that it was hard to get good measurements because it was hard to get a good seal at the cap.

Table 6.3: Calculated and measured values for impedance and admittance as functions of monopole height. Radius = 0.0509λ. Coax outer/inner = b/a = 1.189.

h_m/λ	Z_{comp}, Ω	Z_{King}, Ω	Y_{comp}, mS	Y_{King}, mS
0.1	3 – *j*27.1	5.9 – *j*37.5	4.04 + *j*36.5	4.1 + *j*26
0.12	4.57 – *j*23.8	8.1 – *j*31.6	7.78 + *j*40.5	7.6 + *j*29.7
0.14	7 – *j*20.8	11 – *j*25.9	14.6 + *j*43.2	13.9 + *j*32.7
0.16	10.4 – *j*18	15.4 – *j*20.7	24 + *j*41.7	23 + *j*31
0.18	15.1 – *j*16.1	20.4 – *j*16.1	31 + *j*33	30 + *j*24
0.2	20.7 – *j*15.8	27.1 – *j*12.4	30.5 + *j*23	30.5 + *j*14

We see that for all but the two tallest cases the measured reactance magnitude is larger than the computed value, as expected from King's remark. The ratio of measured to computed resistance starts out at 2:1 and decreases to 1.3:1 as resonant length is approached. My examination of the data in [1] and [2] shows that the resistance also decreases as the coax fit is tighter, although it isn't possible to quantify the trend. On the other hand, the conductances track quite closely, and the susceptances have a nearly constant difference. King and many others of the period talk about a correction term, which accounts for the difference between their theoretical values and the measured values of susceptance. At any rate, the admittance results give us some confidence in our numerical model.

The antennas represented in Table 6.3 are not really electrically small, because a monopole of even 0.1λ is equivalent to a free-space dipole of 0.2λ. So, taking our modeling principles down to h_m = 0.05λ, we find the following results. The Q is from equation (3.21).

Table 6.4: Impedance and Q for an EDM with h_m = 0.05λ.

a/λ	Z, Ω	Y, mS	Q
0.0125	0.5 – *j*107	0.044 + *j*9.3	221.6
0.025	0.6 – *j*64	0.15 + *j*16	112.4
0.05	0.86 – *j*37	0.63 + *j*27	45.4

The Q values are much better than those predicted by Q_w in Table 6.1. The efficiencies are nearly 100%, but the tuning coil loss, assuming Q_{coil} = 200, would run to 0.5Ω for the thinner case, making the combined efficiency about 50%. For a = 0.05λ, the efficiency would be about 82%.

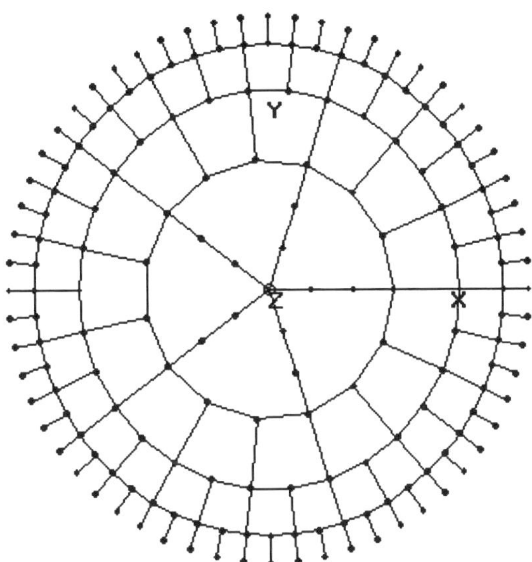

Figure 6.3: Top view of a CDM model. The number of angle steps in the outer ring is 60, then 30, 15, and 5. The dots mark wire ends, and the wires in the inner ring have equal lengths but decreasing radius toward the center.

Modeling the CDM was not as easy as it appeared at first. Using the EX 0 source model gave APG values well below 2, and playing with the drive-wire segments didn't improve this. I have found that the EX 5 source model can be tuned with the segment length/radius to give APG = 2 as close as you are willing to fiddle. For the case using $a = 0.0286\lambda$ of [1], and the gridding described above, a segment length of about $0.001\lambda = 1$ mλ and a radius about 0.29 mλ works. However, using the simple disc gridding of Figure 6.2 produces a very thin wire radius for the simulated disc at its center. This causes the ratio of drive-wire radius to disc-wire radius to be very large compared to the 5:1 NEC guideline. A solution to this problem is shown in Figure 6.3, in which the number of angle steps is progressively reduced in successive rings. This provides fewer inner wires that have to have larger radius values to satisfy the equal-area rule. An algorithm to do this is encoded in the function symdisc2(). The program that generates the NEC wire list is symcyl2.exe, which uses symcyl.in for its input parameters. Table 6.5 shows the disc model values for Figure 6.3.

Table 6.5: Gridding parameters for the disc of Figure 6.3 and the CDM with a = 28.6 mλ.

AngleStep,deg.	6	12	24	72	72	72
WireLengths,mλ	2.85	4.88	7.23	4.55	4.55	4.55
WireRadius, mλ	0.453	0.777	1.15	2.27	1.36	0.455

The algorithm starts with a number of angle steps in the outer ring near the number specified in symcyl.in, then divides by 2, or 3, to get the number of angle steps for the next ring. The radial wire length is increased to equal the average arc length of the cell. The radius of the inner ring is evenly divided to make up the required number of radial steps, six in this case. The wire radii decrease because the area each wire has to simulate decreases going toward the center.

I varied the feed wire length from 3 to 40 mλ, as shown in Table 6.6. The result for 5 mλ is quite close to Brown and Woodward's reactance value, but this is probably just coincidence. I think the lesson to take from these results is that the capacitance between the antenna base and the ground plane shunts the current away from most of the cylinder. Note that the second row in Table 6.6 is comparable in impedance and Q to the second row of Table 6.4. The entries for longer feed wires have progressively better impedance and Q values. The radiation is mostly leaving the feed wire and the function of the remaining cylinder is simply to capacitively load it. One could take the 40-mλ case as pointing towards volume-loading, although this was not the path of discovery. The radiation resistances for uniform-current monopoles are 2.53Ω for $h_m = 40$ mλ and 3.95Ω for $h_{\text{m}} = 50$ mλ. The 40-mλ drive wire in Table 6.6 is doing pretty well, with a small contribution from the edge of the top-loading cylinder.

Table 6.6: Impedance and Q for the Brown and Woodward CDM, as a function of drive wire length l. h_m = 50 mλ, a = 28.6 mλ. Wire radius a_w is also listed. In all cases, segment length is 1 mλ and the EX 5 source is applied at the second segment above ground.

l, mλ	a_w, mλ	Z, Ω	Y, mS	Q
3	0.3069	$0.3 - j35$	$0.24 + j29$	138
5	0.2966	$0.5 - j43.44$	$0.26 + j23$	108
10	0.2862	$0.94 - j54.54$	$0.32 + j18$	80
20	0.2807	$1.67 - j63.3$	$0.42 + j16$	62
40	0.279	$2.93 - j79$	$0.47 + j13$	53

6.2 Top Loading

6.2.1 The Inverted-L

The inverted-L is the simplest case of top loading and it is used occasionally for this reason. It is illustrated in Figure 6.4, with labels to facilitate the discussion. By using the theory of images and some generally-available results for parts of the image-theory structure, we can get analytical approximations for the input impedance. Figure 6.5 shows the image-theory version of the antenna and Figure 6.6 shows an illustration of its current distribution. I assume that the total wire length is short enough to allow me to approximate the current as a triangle function. Since the current doesn't go to zero until the end of the wire, the average value on the vertical section is raised from the minimum of half of the input value. The second benefit of this arrangement is that the radial wire increases the capacitance, which decreases the reactance compared to the simple monopole.

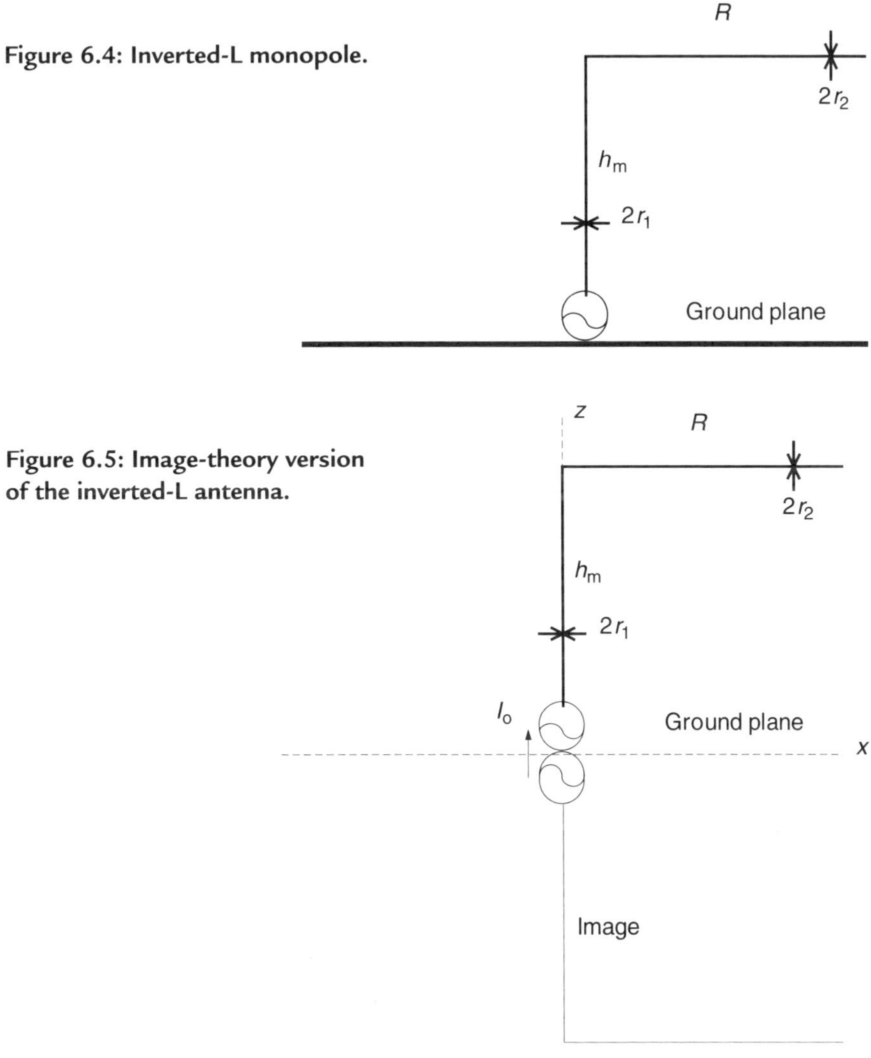

Figure 6.4: Inverted-L monopole.

Figure 6.5: Image-theory version of the inverted-L antenna.

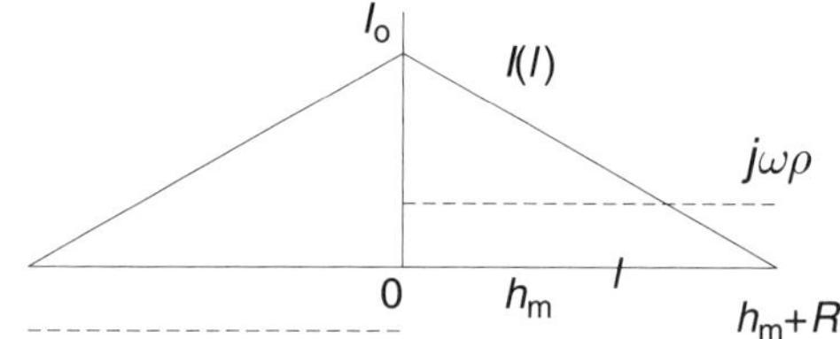

Figure 6.6: Current and charge density on the antenna and its image.

The resistance part of the input impedance is easy; we just need to calculate the average current on the vertical section and use equation (2.39). Figure 6.6 shows the current as a function of length l along the wire. On the vertical section, $l = z$, and on the upper radial, $l = h_m + x$. The slope on the right-hand side of Figure 6.6 is $m = -I_o/(h_m + R)$, so the expression for current is

$$I(l) = I_o + ml \tag{6.6}$$

The average on the vertical section is found by integrating from 0 to h_m.

$$I_{avg} = \frac{1}{h_m}\int_0^{h_m}(I_o + mz)\,dz = \frac{1}{h_m}\left(I_o h_m + m\frac{h_m^2}{2}\right) = I_o\frac{R + h_m/2}{R + h_m} = \alpha I_o \tag{6.7}$$

Adapting (2.39) for a monopole, we have

$$R_r = 160\left(\frac{\pi h_m}{\lambda}\cdot\frac{R + h_m/2}{R + h_m}\right)^2 \tag{6.8}$$

You can see that $\alpha \to 1$ as R becomes large compared to h_m. Reaching $\alpha = 1$ isn't likely in practice, but having R 3 or 4 times h_m is quite practical.

To get at the input reactance, I used the stored energy approach to capacitance. If the charge on a capacitance C is $q(t) = Q_o\sin(\omega t)$, the instantaneous stored energy is $w_e = q^2/(2C)$. The mean energy is the average over a cycle, $W_e = Q_o^2/(4C)$. Charge is the integral of current into the capacitor, so for a current of amplitude I_o the mean stored energy is:

$$W_e = \frac{I_o^2}{4\omega^2 C} \tag{6.9}$$

The energy stored in a continuous charge density distribution is the limiting case of that for discrete charge assemblies, [3, pp. 211–214],

$$\begin{aligned} w_e &= \frac{1}{2}\sum_j q_j V_j = \frac{1}{2}\sum_j q_j \sum_i \frac{q_i}{4\pi\epsilon_o R_{ij}} \\ &= \frac{1}{8\pi\epsilon_o}\iint\frac{\rho(z)\rho(z')}{R(z,z')}dz\,dz' \end{aligned} \tag{6.10}$$

In the integrand of (6.10), the $\rho(z)$ is a line charge density, *C/m*, and the R_{ij} and $R(z,z')$ are the distances between an element of charge producing a potential field and an element of charge pushed into position (given potential energy, worked on) against the field of all the other charges. The continuity equation that relates current and line charge density in phasor form is:

$$\frac{dI}{dl} + j\omega\rho = 0 \tag{6.11}$$

Since the slope of I is constant, so is the charge density. The part of (6.10) that gives the instantaneous energy from a line charge distribution can now be written directly as a mean energy for sinusoids.

$$W_e = \frac{m^2}{16\pi \in_o \omega^2} \iint \frac{dzdz'}{R(z,z')} = \frac{m^2}{16\pi \in_o \omega^2} I_R \tag{6.12}$$

where, again, m is the current slope.

I conceptually divided the antenna and its image into two parts. The z axis part is a dipole, and the radial and its image are a parallel-wire transmission line section. The energy stored on the dipole will have the same form as that stored in an open-ended dipole—only the charge density is different. Therefore, I can use our knowledge of the capacitance of the open-ended dipole to find the value of the double integral in (6.12). Adapting (2.45) to the present situation's notation, the capacitance of the open-ended dipole is:

$$C_d = \in_o \pi h_m \Big/ \left[\ln\left(\frac{h_m}{r_1}\right) - 1 \right] \tag{6.13}$$

The current slope in the open-ended dipole is I_o/h_m. Equating the two energy expressions, (6.9) and (6.12) for the open-ended dipole gives us I_R in terms of C_d.

$$I_R = \frac{4\pi \in_o h_m^2}{C_d} = 4h_m \left[\ln\left(\frac{h_m}{r_1}\right) - 1 \right] \tag{6.14}$$

Now I can use (6.12) to find the mean stored energy of the dipole part of the inverted-L and its image, W_{di}.

$$W_{\text{di}} = \frac{I_o^2 h_m}{4\pi \in_o (h_m + R)^2} \left[\ln\left(\frac{h_m}{r_1}\right) - 1 \right] \tag{6.15}$$

The parallel-wire transmission line part has a capacitance per unit length, adapted from [3, pp. 362–365], given by

$$C_{\text{pu}} = \pi \epsilon_o \Big/ \cosh^{-1}\left(\frac{h_m}{r_2}\right) \approx \pi \epsilon_o \Big/ \ln\left(\frac{2h_m}{r_2}\right), \quad h_m \gg r_2 \tag{6.16}$$

Again, the charge density is $\rho = I_o/(j\omega(h_m + R))$, the total charge is $R\rho$, the total capacitance is RC_{pu} so the mean energy stored in the radial and its image is:

$$W_{\text{pw}} = \frac{R\rho^2}{4C_{\text{pu}}} = \frac{I_o^2 R}{4\pi \epsilon_o \omega^2 (h_m + R)^2} \ln\left(\frac{2h_m}{r_2}\right) \tag{6.17}$$

Adding (6.15) and (6.17) gives us the total mean stored energy. Using (6.9) gives us the equivalent circuit capacitance for the dipole, and multiplying by 2 gives us the capacitance for the monopole and its actual source.

$$C_{IL} = 2\pi \epsilon_o (h_m + R)^2 \Big/ \left\{ h_m \left[\ln\left(\frac{h_m}{r_1}\right) - 1 \right] + R \ln\left(\frac{2h_m}{r_2}\right) \right\} \tag{6.18}$$

The result could be regarded as two capacitances in series. Equations (6.8) and (6.18) reflect the earlier comments about the effects of the radial wire length.

Example 6.1 **Inverted-L on a PEC plane**

I chose to examine an inverted-L that fits in the same space as one of the fat monopoles of the last section. I made $h_m = 50$ mm, $R = 100$ mm, and the wire radius 1 mm for both. I wrote a MATLAB function to generate R_{rad} and X_{in} values using (6.8) and (6.18), and a NEC wire list. The results are shown in Table 6.7.

Table 6.7: Inverted-L impedance as a function of frequency. Z_a is from (6.8) and (6.18). Z_{nec} is from a model with 5 segments on the vertical and 10 segments on the radial wire. APG = 2.015

f, MHz	*λ, m*	$\frac{h_m + R}{\lambda}$	Z_a	Z_{nec}
50	6	0.025	0.076–*j*1543	0.077–*j*1560
100	3	0.05	0.305 –*j*772	0.312 –*j*761
150	2	0.075	0.685 –*j*514	0.713 –*j*486
200	1.5	0.1	1.219 –*j*386	1.297 –*j*341
250	1.2	0.125	1.904 –*j*309	2.092 –*j*248
300	1	0.15	2.742 –*j*257	3.136 –*j*181

The resistance values are quite close, with the NEC results being slightly higher and the difference growing with frequency. This probably reflects the radiation from the radial wire, neglected in the above analysis. At 300 MHz, the total wire length is 0.15λ, which diminishes the accuracy of the triangle-current approximation. Note that the resistance is better and the Q is worse than the corresponding EDM in Table 6.4, $a/\lambda = 0.05$.

Example 6.2 **Inverted-L on a Box**

For this example, I hypothesized a 915-MHz band radio with space constraints. The radio box is assumed to be 10 mm × 40 mm × 80 mm, and the antenna is to fit inside the plastic package. I chose a height of 10 mm, located the antenna on a short edge, and ran the radial out the 40-mm width of the box. Figure 6.7 illustrates a wire list generated by LonBox.exe. All segments are 2.5 mm long. The vertical portion of the antenna has 4 segments, the radial wire has 16, and the box has 4 by 16 by 32 segments. The equal-area rule gave the box segments a radius just under 0.4 mm. I used an EX 5 source and tuned the antenna wire radius to get APG = 1.000.., which came to 0.67 mm. The impedance was $10.6 - j125\Omega$ and Q was about 18. Notice that the Q is higher than $|X|/R$, which indicates that the impedance is moving up a resonance curve. A series resonating coil with $Q = 200$ would have a loss resistance of about 0.6Ω. Even with an L-section matching network, the efficiency should be better than 90%.

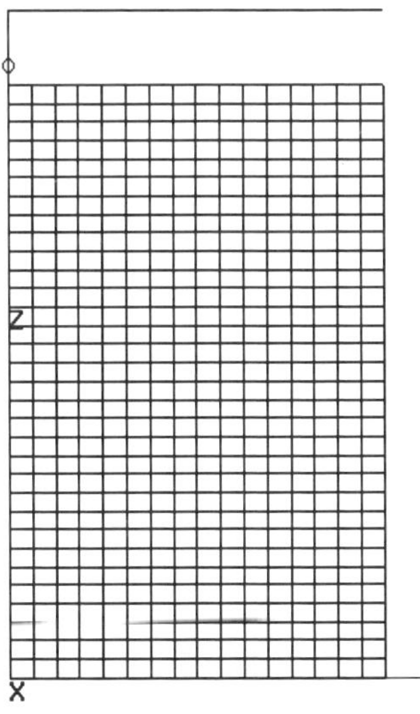

Figure 6.7: Side view of wire model for an inverted-L antenna on a radio box.

I also tried other modeling choices. Replacing the EX 5 with an EX 0 source gave $Z_{in} = 11.6 - j132\Omega$, with an APG = 1.016. Reducing the antenna wire radius to 0.4 mm with the EX 0- source gave $Z_{in} = 12.15 - j157\Omega$ with APG = 0.977. Higher reactance is to be expected from the thinner wire, as long as capacitance dominates inductance.

For a size comparison $\lambda/4 = 82$ mm. I modeled a whip at the same antenna position, and tuned its length to get $R_{in} = 50\Omega$. 55 mm height with the same 0.67 mm radius gave $Z_{in} = 53.66 - j96\Omega$ with APG = 0.957. The designer's choice is to use a whip that sticks out of the package and one inductor, or an internal antenna with one inductor and one capacitor.

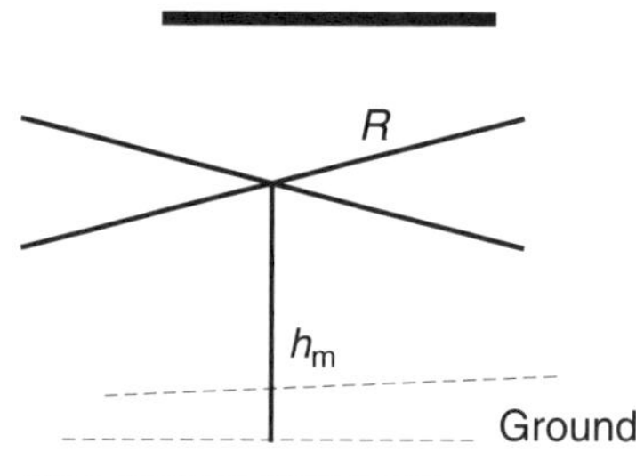

Figure 6.8: Illustration of monopole with four top-loading radial wires.

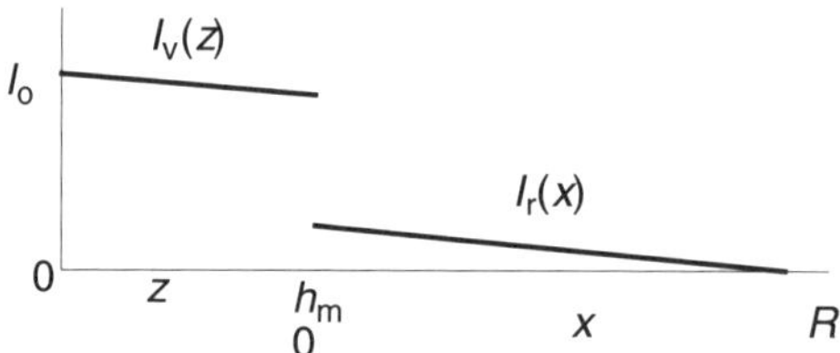

Figure 6.9: Illustration of currents for the antenna of Figure 6.8.

6.2.2 Top-Loading with Radials

Figure 6.8 illustrates a monopole with four radial wires, and Figure 6.9 depicts the current along the vertical and one radial. We can extend the analysis of the inverted-L, a one-radial case, to multiple radials without much difficulty. For this purpose, I will discuss only the case where all wires have the same radius, as this makes the charge density at the junction a simple issue. The currents and charge densities at the junction must meet two conditions: the radial currents have to add up to the incoming vertical current, and the line charge densities have to be equal. Let the current on the vertical wire be:

$$I_v(z) = I_o - m_z z, \quad 0 \le z \le h_m \tag{6.19}$$

and the current along one radial be:

$$I_r(x) = I_{ro} - m_x x, \quad 0 \le x \le R \tag{6.20}$$

Since $I_r(R)$ must be zero, $m_x = I_{ro}/R$. For n radials, $nI_{ro} = I_v(h_m) = I_o - m_z h_m$. Equal line charge densities at the junction means the current derivatives have to be equal. Then $m_x = m_z$ and $nRm_z = I_o - m_z h_m$,

$$m_z = I_o/(nR + h_m) \tag{6.21}$$

The effect of the radials seems to be the same as if one very long wire were used, as long as it's still electrically short. Using the same reasoning as for (6.8) and (6.18), we arrive at:

$$R_{\text{rad}} = 160\left(\frac{\pi h_m}{\lambda} \cdot \frac{nR + h_m/2}{nR + h_m}\right)^2 \tag{6.22}$$

$$X = \frac{-30\lambda}{\pi(nR + h_m)^2}\left\{h_m\left[\ln\left(\frac{h_m}{r_w}\right) - 1\right] + nR\ln\left(\frac{2h_m}{r_w}\right)\right\} \tag{6.23}$$

The common wire radius is r_w. If the radii were unequal, the line charge densities would not be continuous at the junction, and a more complex relation would have to be used, as in NEC.

One can use an expression for magnetic stored energy, [4, eq.(16)], to derive a term for the inductive reactance of the vertical section. Even at 1/20λ this is a significant quantity compared to the other reactance components.

$$X_L = 60\beta h_m\left(\frac{nR + h_m/2}{nR + h_m}\right)^2\left[\ln\left(\frac{h_m}{r_w}\right) - 1\right] \tag{6.24}$$

The common thread in all of these results is the idea of calculating stored energy for either uniform (constant in space) charge density or uniform current density and relating that energy to the terminal current to get a reactance reflected to the input terminals. Even though the equations are not terribly accurate, and overstate the benefit of loading, they illustrate some pertinent physical facts. As the loading increases, the average vertical current increases towards the terminal value. This causes the charge density to decrease, which causes a shift from capacitive toward inductive reactance for the vertical element. Increasing the number of radials decreases their reactance magnitude, making series resonance possible.

Example 6.3 **Radial-Loaded Monopoles**

I have written models and model-generators for both monopoles and verticals loaded at both ends. As an example to illustrate some of the things we've been discussing, consider a 50-mm vertical with 50-mm radials over a PEC plane. The wire radius is 0.5 mm in all cases. 50 MHz and 300 MHz are the test frequencies, and $\lambda = 6$ m and 1 m.

Table 6.8: Impedance as a function of the number of radials. Vertical and radial wires are 50 mm long, 1-mm diameter. The NEC models have 20 segments per wire.

	50 MHz		*300 MHz*	
n	Z_a	Z_{nec}	Z_a	Z_{nec}
2	$0.076 - j1800$	$0.066 - j1813$	$2.74 - j254$	$2.67 - j246$
4	$0.089 - j1128$	$0.079 - j1279$	$3.2 - j134$	$3.18 - j145$
8	$0.098 - j641$	$0.088 - j941$	$3.52 - j48$	$3.52 - j81$
16	$0.103 - j340$	$0.093 - j752$	$3.72 + j5.6$	$3.71 - j45$

The reason for showing the 50-MHz results is that there is no significant inductance component at this frequency. The reactance magnitude for the analytical impedance drops almost by half for each doubling of the radials, but the computed result does not. This shows that something more is going on than just paralleling equal reactances. At 300 MHz, the inductance is important, and the computed reactance magnitude here does drop by almost half per radial doubling, and that for the analytical expressions moves even faster.

I copied the charges and currents for the two 4-radial cases into .dat files, loaded them into MATLAB variables and plotted them in Figures 6.10 and 6.11. The source voltage was 100. It is apparent that the charge density is not constant in space, nor does any variable fit the popular sinusoidal assumption. The average charge densities are shown in Table 6.9.

Table 6.9: Average charge densities for Figures 6.10 and 6.11. The units are nC/m.

	50 MHz	*300 MHz*
$q(z)$	1.1	1.26
$q(x)$	0.94	1.5

It appears that charge is shifted to the radials as frequency goes up, another indication that the reactance of the vertical is less capacitive. For the 16-radial, 300-MHz case, the Q is 29.5. The equivalent L and C are 45.7 nH and 4 pF.

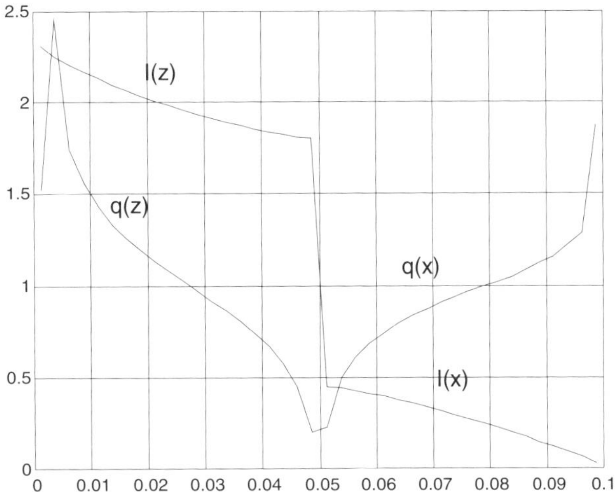

Figure 6.10: Current and line charge density for 50 MHz. *q* is in nC/m and the current is in A, and multiplied by 30 to fill the scale.

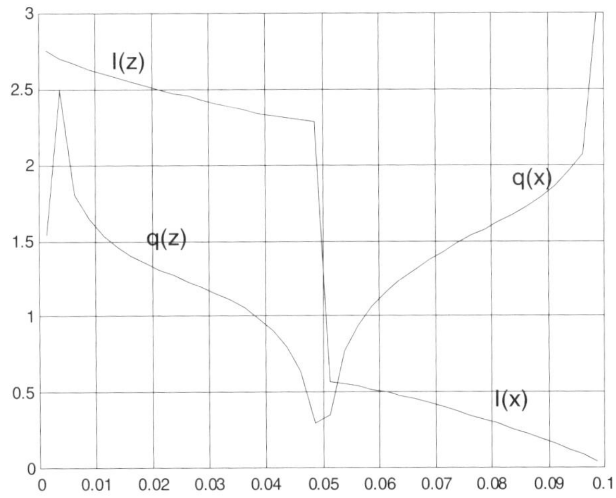

Figure 6.11: Current and line charge density for 300 MHz. *q* is in nC/m, and the current is A, multiplied by 4 to fill the scale.

6.2.3 Volume Loading

Volume loading means topping the vertical wire with a conducting three-dimensional shape instead of the two-dimensional disc or wire-radial imitation disc. If the body has the right shape, the effect is to produce a charge distribution that peaks strongly at the body's outside edge, maximizing the capacitance in the allowed space. The effect was discovered by numerical experiment in the

'90s, and the early work is described in [5-6]. Figure 6.12 is an illustration of the profile of a volume-loaded dipole. The arrows show the reference directions of current flow around the profile. Current travels along the central radiating wire, then spreads out when it reaches the end body. At resonance, the current is entirely real, with respect to the applied voltage, and nearly uniform along the wire. It decreases gradually going out toward the outer rim, drops sharply at the rim, and goes gradually to zero at the center of the outside face. Since charge density is proportional to the space derivative of current, the charge density peaks at the rim. Basically, resonance is produced by balancing the inductance of the wire against the capacitance of the end bodies. From the last section, it is clear that radial-loading can be used to induce resonance as well, but it does not produce as much capacitance in the given space as volume loading, so the designer either has to use more space for the radials, thinner wire for the radiator, or more wire coiled to raise the inductance and fit the space allowed. The design process for volume-loading is entirely empirical; no analytical help is available at this time.

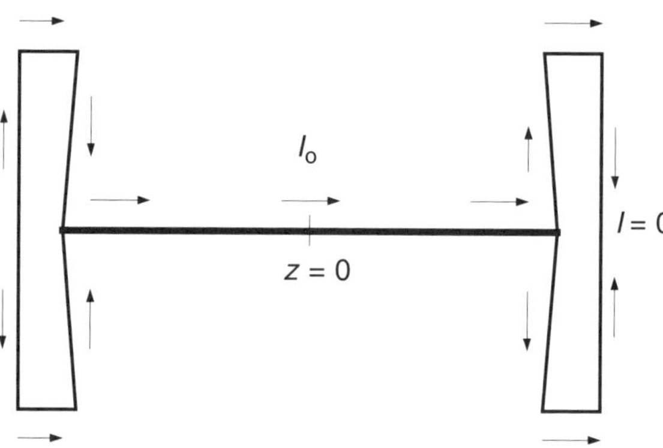

Figure 6.12: Illustration of a volume-loaded dipole.

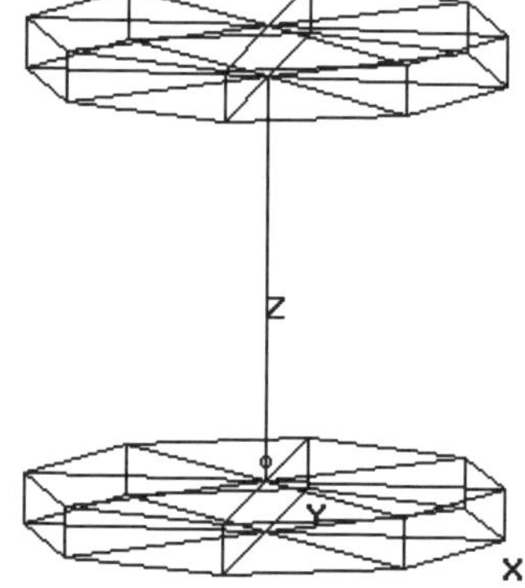

Figure 6.13: Volume-loaded dipole designed using framework end bodies.

Example 6.4 **A Volume-Loaded Dipole at 40 m**

Figure 6.13 shows the picture of the NEC wire list for a dipole that is 4 m tall by 4 m wide. The end bodies are octagonal frames made from ½" copper water pipe. The central wire radius is 1.6 mm, approximately AWG #8, and it is 3.6 m long. In free space, the dipole resonates at 7.1525 MHz, 42 m, with a resistance of 6.13Ω. The uniform-current radiation resistance for 3.6 m is 5.8Ω, so there is a little help from the verticals of the end frames. The Q is 38.36. The equivalent L and C are 5.23 μH and 94.7 pF. In terms of wavelength, this structure is nearly the same size as the equivalent dipole for Example 6.3. The Q is higher for the present case than the 16-radial case, but the really striking difference is the effective capacitance. The 16-radial case dipole equivalent has an effective capacitance of 2 pF, 47 times less than the 8-radial framework volume load. For a direct comparison, the computed values for the 8-radial dipole equivalent from Example 6.3 are $Q = 35$, $L = 87$ nH, and $C = 1.63$ pF. The small inductance values are at least in part due to the wire radius being $\lambda/2000$ for Example 6.3 and $\lambda/26{,}250$ in the present case.

Figure 6.14 shows the current and charge distribution along the profile of the antenna. The current is real, as expected at resonance, and the charge density is imaginary, as required by the continuity equation. The straight-wire part of

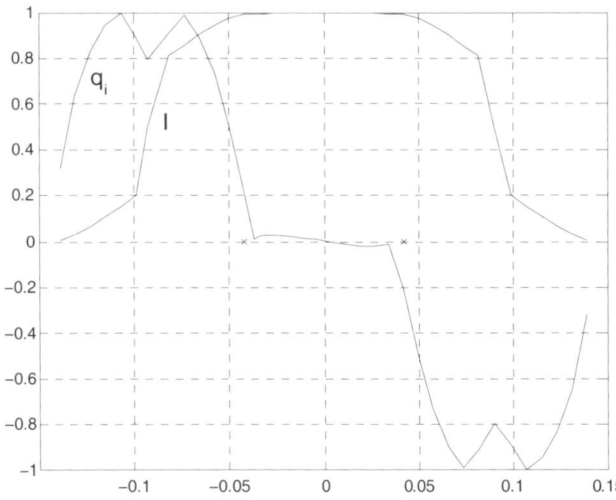

Figure 6.14: Normalized current and line charge density for the dipole of Figure 6.13. The distance is in wavelengths. The plot is of total value at a cross-section on the profile. The *x*s mark the ends of the straight-wire radiator. The total current or charge at a point on the frame's profile is found by taking the value for one segment and multiplying by 8. There are 38 segments in the profile.

the antenna runs between the x marks, which are at $\pm 0.043\lambda$. The charge peaks are on the outermost segments of the frame's radials. The vertical pipe between the two layers of a frame is represented only by one segment, hence the choppy appearance of the charge plot. Reading the plot's horizontal axis, 0 is at the dipole's center, and going to the right moves along the wire until you reach the x. Since 2 m is about 0.05λ, the next 0.05λ corresponds to moving out along the radials. As mentioned, the scale moves to one point on the rim, and then the final 0.05λ corresponds to moving along the radials from the rim to the outside center. As in Example 6.3, you can see that the distributions are not sinusoidal functions of position.

6.3 Coil Loading

Top-loads are not very aerodynamic, and they are generally not self-supporting. Above 800 MHz it's easy to integrate them into the packaging, but below 30 MHz the smallest mobile platform that can use them is a ship. One solution that's been around for over 50 years is to put a coil partway up the monopole. The coil is adjusted to series resonate the portion of the antenna above it, which makes the current in the lower section nearly uniform. In effect, the coil and the upper antenna section are a top load for the lower antenna section. See Figure 6.15 for an illustration. This raises the radiation resistance and improves system efficiency over that of a simple monopole with a series coil at the base. The improvement is limited by the fact that the section of antenna above the coil has less capacitance than the unloaded antenna, so the coil inductance has to be larger than one for base tuning, and so the coil loss resistance is larger. The earliest analysis I know of was published by J. S. Belrose in 1953 [7] and was based on an assumed linear current distribution on each part of the antenna. A more elaborate analysis was published by C. W. Harrison [8] in 1963 based on general sinusoidal current models. As noted in Chapters 2 and 3, the plain monopole is not itself inefficient; its low radiation resistance combined with the fact that a tuning or matching coil will have a lower Q than the antenna makes the system very inefficient. In the following three examples, I show the properties of center-coil loaded monopoles and compare NEC results with Harrison's analytical ones.

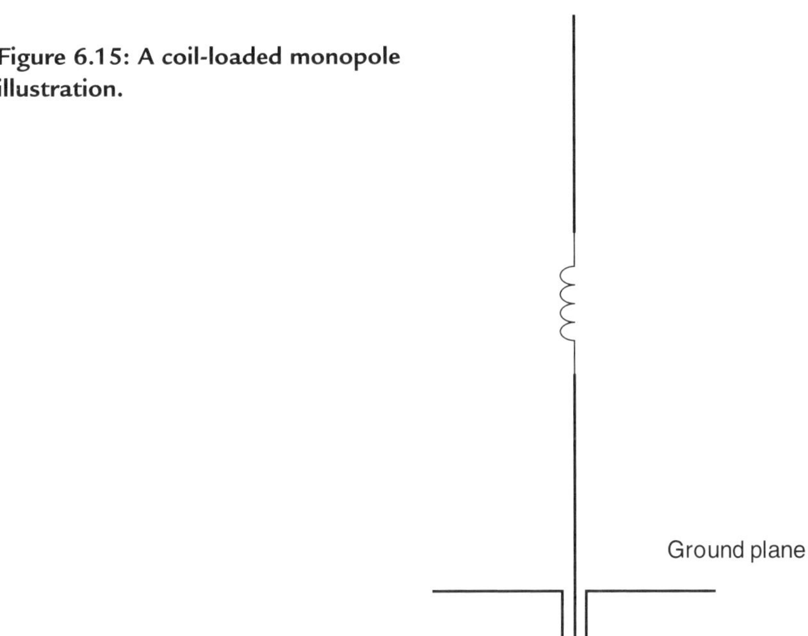

Figure 6.15: A coil-loaded monopole illustration.

Example 6.5 ***A Very Short Center-Loaded Monopole***

Harrison's paper is a summary of a report done for Sandia Corp. in which he apparently gave many tables of design calculations. The paper includes two tables that give the variation of input impedance with loading coil L and Q for a fixed size (Table I), and the value of resonating X and input resistance for various size values and coil placements (Table II). The increase in resistance at the base with the decrease in Q shows how inefficient the system is with even very good coils. His calculations in Table I are for $\beta h_m = 0.2$, centered coil, and $r_w/\lambda = 0.000122895$ so I call this the Harrison monopole and use it for this example.

At 300 MHz, with $\lambda = 1$ m, $h_m = 31.83$ mm and $r_w = 0.122895$ mm. I modeled the monopole with 41 segments so I could put the loading coil at the exact center in segment 21. Listing 6.1 shows the final version of the model. Harrison gives the no-coil input impedance as $0.385 - j1350\Omega$, the resonating coil reactance for center-loading as 2448Ω, and the input resistance at resonance as 1.185Ω. The reactance corresponds to $L = 1.2987$ µH. The simulation gave a no-coil input impedance of $0.38 - j1343\Omega$, pretty close. However, I had to tune the coil to 1.33354 µH to $Z_{in} = 0.997 - j0.007\Omega$, not so close. You can see from the number of digits in L that tuning was quite fussy. This implies touchy tuning in the physical case as well. Figure 6.16 shows the current and charge density distributions as calculated by NEC for no losses. From the current plot, assuming a constant value below the coil and a straight-line run down to zero above the coil doesn't seem too bad. In the no-coil case, the radiation resistance is $R = 40(\pi h_m/\lambda)^2 = 0.4\Omega$. With the coil in the center and making the straight-line approximations, the average current is ¾ I_o, so $R = 40(\pi(3/2)h_m/\lambda)^2 = 0.9\Omega$. Not too bad.

Listing 6.1: The Harrison Monopole

```
CM Coil-loaded monopole.
CE
GW 1 41 0 0 0 0 0 31.83 0.122895
GS 0 0 0.001
GE 1
GN 1
LD 0 1 21 21 5.5858 1.33347e-6 0
LD 5 0 0 0 26e6
EX 0 1 1 1 100 0 1
FR 0 1 0 0 300 2
PT -1
RP 0 46 46 1002 0 0 2 2
EN
```

The input voltage was arbitrarily set at 100—rather excessive, unless we are talking about an LF beacon, because it implies 10 kW input power. A point that Harrison makes is that the coil voltage is higher, frequently much higher, than the input voltage. The coil voltage in this case is 2514Ω × 111.46A = 280.25 kV. V_{coil}/V_{in} = 2802, a high ratio, whatever the input voltage might be.

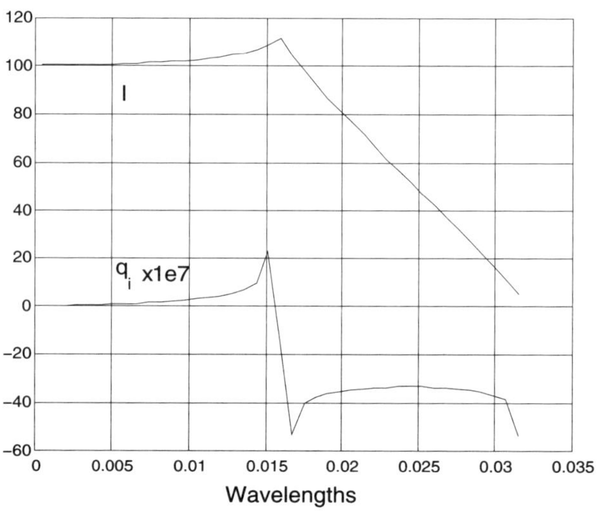

Figure 6.16: Current, A, and line charge density, C/m, along the Harrison coil-loaded dipole, no-loss case.

Next, let's look at the effect of losses. Assuming a wire conductivity of 26 MS/m caused the efficiency to go to 83%, and required retuning the coil to 1.33346 to get $Z_{in} = 1.2 + j0.008\Omega$. Now assume a coil Q of 200 and add 12.57Ω to the LD 0 command. Retuning is again needed; $L = 1.333504$, $Z_{in} = 1\ 6.7 - j0.002\Omega$, with an efficiency of 6%. Suppose we had just used a coil at the base? A base coil of 712.44 nH with a Q of 200 gives $Z_{in} = 7.18 - j0.0026\Omega$, with an efficiency of 5.3%. If coils with a Q of 450 can be built, base-tuning gives $Z_{in} = 3.45 - j0.0026\Omega$ and an efficiency of 11%. Center-loading with $Q = 450$ gives $Z_{in} = 8.1\Omega$ and an efficiency of 12.3%. The center-loading cases don't seem much more efficient, but the input impedance in any case still has to be transformed to 50Ω by an L-section. The higher impedance of the center-loaded antenna means the coil in the L-section will be smaller and degrade system efficiency less.

Example 6.6 **The Harrison Monoopole at 3 MHz over Real Ground**

In this example we move a little closer to a real application. I scaled the Harrison monopole for 3 MHz, which means the height and wire radius are multiplied by 100 from the last example. A stationary vertical antenna on real ground could be driven against a ground stake, but this puts most of your power into the ground if the antenna is short. Usually such an antenna is surrounded by wire laid on or in the ground to reduce the amount of RF power going into the ground. For the present case, I chose to put down 8 radials, each 5 m long and having a wire radius of 3.2 mm. The radius is much larger than would be used in practice, but it meets the NEC junction guidelines with respect to the vertical wire radius without making up some unphysical transition wires. The radials and antenna are 0.1 m (about 4") above the ground, both because NEC2 can't handle radials in the ground and because numerical and physical experiments have shown that elevated radials produce less loss. The radials are also electrically short because I imagine them fitting in a typical house lot backyard. The model's final version is given in Listing 6.2.

Listing 6.2: The Harrison monopole and radials over real ground.

Note that the coil is represented by an LD 4 command. This command specifies the load as resistance and reactance. There is no frequency dependence in this command.

```
CM Coil-loaded monopole and radials over real grounad.
CE
GW 1 41 0 0 0.1 0 0 3.283 0.0122895
GW 2 64 0 0 0.1 5 0 0.1 0.0032
GM 1 7 0 0 45 0 0 0 2
GE 1
GN 2 0 0 0 13 0.005
LD 4 1 21 21 5.738 2582.25
EX 0 1 2 1 100 0 1
FR 0 1 0 0 3 0.1
PT -1
RP 0 46 46 1002 0 0 2 2
EN
```

Without loading or tuning and no wire loss, $Z_{in} = 1.894 - j1457\Omega$ and APG = 0.1805. Since we are operating over a ground plane, APG = $2P_{rad}/P_{in}$, where P_{rad} is the power radiated into the air above the plane. Therefore, $R_{rad} = R_{in}\text{APG}/2$. In the present case, $R_{rad} = 0.1752\Omega$. The remaining resistance is all due to the power pumped into the ground, often called the *ground resistance*. In this case, $R_{gnd} = R_{in} - R_{rad} = 1.72\Omega$. Like the radiation resistance and the sky-wave, ground resistance is a terminal reaction caused by electromagnetic radiation into the ground.

From the previous example and other discussions, it is apparent that resonating the antenna at any point makes it resonant everywhere. To find out how much reactance I need to add at any segment, I put a source in that segment in the model and let NEC find the impedance it sees. I did this for segment 21, and the impedance was $3.73 - j2582.23\Omega$. I tested both base-tuning and center-loading with coils having Q of 200 and 450. The results are collected in Table 6.10.

Table 6.10: Base and center tuning the 3-MHz vertical over real ground. σ = 5 mS/m, ϵ_r = 13. *Eff* = structure efficiency, which accounts for power lost in the wires (none, in this case) and loads. APG = average power gain.

Base Tuning

Q	Z_{in}	*Eff. %*	*APG/2*
200	9.18	20.63	0.01862
450	5.132	36.9	0.0333

Center Loading

Q	Z_{in}	*Eff. %*	*APG/2*
200	21.54	21.59	0.02007
450	12.25	38.74	0.0353

APG/2 is effectively the radiation efficiency, so it is the most important number to watch. Again, there is very little improvement in going from base-tuning to center-loading, and both are worse than no tuning. However, you can't operate without tuning because the reactance is much too high.

In the previous example there were only structure losses, so efficiency/100 and APG/2 should be equal. Define E = (Structure Efficiency)/100, and A = APG/2 is the radiation efficiency in the presence of a ground plane. In general, $R_{\text{loss}} = (1 - E)R_{\text{in}}$, $R_{\text{rad}} = R_{\text{in}}A$, and $R_{\text{gnd}} = R_{\text{in}} - R_{\text{loss}} - R_{\text{rad}} = R_{\text{in}}(E - A)$. Doing the calculations for three cases yields the following table.

Table 6.11: Terminal resistance components.
Case 1. No tuning.
Case 2. Base tuning, *Q* = 450.
Case 3. Center-loading, *Q* = 450.

Case	R_{in}	R_{rad}	R_{loss}	R_{gnd}
1	1.894	0.171	0	1.723
2	5.132	0.171	3.238	1.723
3	12.25	0.432	7.504	4.313

The current distribution in Cases 1 and 2 is the same, so the radiated power into the sky and the ground should be the same. The only difference then is the coil loss, which comes out right. In Case 3 the current distribution has the increased average value we expect from center-loading, so the resistances that represent both radiated powers go up. The loading coil's resistance is 5.738Ω. The power dissipated in R_{loss} must be the same as the coil's power loss, so we can find the coil's current as $I_{\text{coil}} = I_{\text{in}}\sqrt{R_{\text{loss}}/R_{\text{coil}}} = 1.144 I_{\text{in}}$. This is what you should see if you change the PT–1 command to PT 0 1 1 41 in the model listing.

Example 6.7 **CB/FM Antenna on a Car**

In this example I present the design process for a two-band car-mounted monopole. To begin, I developed code to generate a model for a hatchback car to be run under 4nec2. The code is in hatch.cpp and compiled into hatch.exe. The input file for hatch.exe is hatch.in, given in Listing 6.3 so you can see the description of the car.

Listing 6.3: hatch.in

```
0.25 1.1 1.4 1
0.2 0.3 1.75 2 3.2
1.025 0.0127  2 2
5 0
2 1.1 0.00635 1.0 11 2 6
30 90 60
0 2
26e6 0.005 13
200 1274.4 13.761
This file, hatch.in is read by hatch.exe to generate
hatch.nec, a 4nec2 file to simulate the hatchback car
sketched below, and a whip antenna with an LC trap
partway up. All elements must be present, whether used
or not. Dimensions are in meters.
```

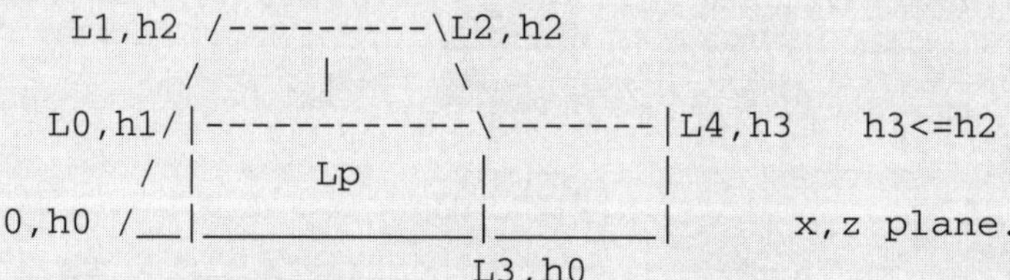


```
Width is W.

By rows:
1. h0, h1, h2, h3.
2. L0, L1, L2, L3, L4.
3. Lp, wire radius for all posts, no. segs. all posts, W.
4. n1, no. vertical segments in main side panel. Sets the
   grid density. EX type, 0 or 5.
5. Antenna: base x,z, wire rad., height, no. segs, source
   seg, load seg. If load seg.=0, no trap is present.
6. Start, stop, step freqs., MHz. If step freq.=0, only
   start freq. is used.
```

```
7. Rad. pattern flag, 1=azimuth pattern. Ground type flag,
   0=free space, 1=perfect, 2=real ground.
8. Wire cond., ground cond., rel. permittivity.
9. Trap coil Q, nH, pF.
```

The car is modeled as being constant width and having a profile defined by straight-line sections. The right side of the car is in the *x,z* plane and headed in the +*x* direction. A drawing of the wire model is shown in Figure 6.17. The grill area, hood, roof, back panel below the window, floor under the back and passengers, and the firewall are also modeled with wire grids. The gridding is as uniform as possible, given the dimensional choices, so all grid spacings are determined by one choice, the number of vertical segments in the door panel.

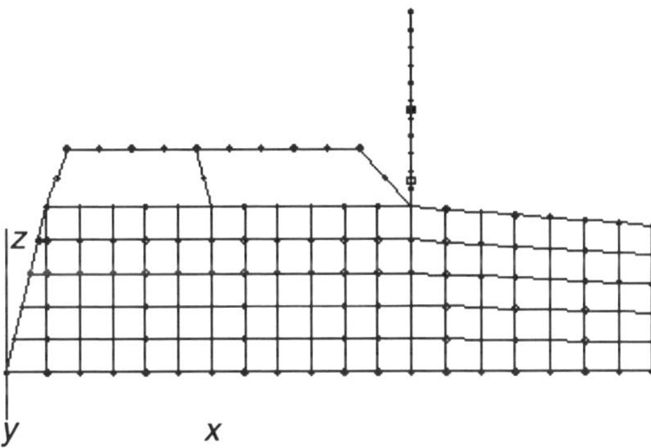

Figure 6.17: Side view of wire model for a hatchback car with a two-band antenna. The segment ends are marked with dots. The antenna has 11 segments, with the EX 0 voltage source in segment 2 and a parallel LC load in segment 6.

The intent of the design is to be able to operate on 30 MHz and 90 MHz. 30 MHz is near both the U. S. Citizens' Band and the amateur radio 10-m band. 90 MHz is close to where most FM public radio stations operate in the U.S. I chose to use a 1-m high antenna, which is a bit long for FM, but short for 30 MHz. To begin the design process, I put the model in free space and tried out a few grid densities and antenna segment densities and positions to get the APG close to 1. I did this at 30 MHz, without an LC trap present. I wound up with an APG = 0.988 and $Z_{in} = 8.77 - j387\Omega$. Next I added the conductivity of aluminum, 26 MS/m, to all the wires. This raised Z_{in} slightly to $8.79 - j387\Omega$, and made the efficiency 99.71%. Over a PEC plane, $Z_{in} = 9.57 - j383\Omega$, APG = 1.85, and the efficiency was 99.75%. Remember that the APG target over a PEC plane is 2E. Close enough.

Moving next to simulating real ground with $\sigma = 5$ mS/m and $\epsilon_r = 13$ (average ground), gives $Z_{in} = 9.46 - j387\Omega$, not much change from the PEC case. The ground parameters make a bigger difference in the lower HF band. The next stage in this example's design process is finding the LC values for the various tunings. I chose the antenna length long enough so it would be inductive at 90 MHz, and of course it's capacitive at 30 MHz. Thus, either base-tuning or center-loading the antenna with an inductive reactance at 30 MHz and a capacitive reactance at 90 MHz should work. A parallel LC circuit (called a trap, in a slightly different application, by radio amateurs) is inductive below its resonant frequency and capacitive above it, while a series LC circuit is capacitive below and inductive above its resonant frequency, so we choose a parallel form. For a parallel form, it's easier to deal with admittance rather than impedance. The susceptance is $B = \omega C - 1/(\omega L)$. Inverting the segment reactances, including signs, at the two frequencies gives the desired values of B, and the two equations in two unknowns (C and $1/L$) are readily solved. See function parallel.m. 4nec2 includes an LD 6 command, which allows you to specify the coil Q instead of calculating the corresponding resistance. This is used in hatch.nec. The results for base tuning and center loading are collected in Table 6.12.

Table 6.12: Design results for the two-band car antenna.
Base tuning: L = 733.7 nH, C = 24.65 pF.
Segment 6 loading: L = 1274 nH, C = 13.764 pF.

Base Tuning, Q = 200 Q = 450

	30 MHz	*90 MHz*	*30 MHz*	*90 MHz*
R_{in}	14.85	73.78	11.8	73.62
E	0.6354	0.9963	0.7968	0.9984
A	0.1867	0.554	0.2341	0.555
R_{rad}	2.77	40.87	2.76	40.86
R_{loss}	5.41	0.27	2.4	0.12
R_{gnd}	6.66	32.63	6.64	32.64

Seg. 6 loading, Q = 200, Q = 450

	30 MHz	*90 MHz*	*30 MHz*	*90 MHz*
R_{in}	24.64	53.3	19.74	53.37
E	0.6419	0.9987	0.8013	0.9983
A	0.1906	0.5502	0.2377	0.5501
R_{rad}	4.696	29.33	4.69	29.36
R_{loss}	8.82	0.07	3.92	0.09
R_{gnd}	11.12	23.9	11.13	23.92

Note that while the resistance at 30 MHz went up to reflect both the coil loss and the increased radiation resistance, it went down at 90 MHz. It is sometimes said that inductive loading lengthens the antenna, while capacitive loading shortens it. The improvement in radiation efficiency by going from base tuning to center loading is tiny, and not worth the extra mechanical fabrication time and cost.

Antenna placement affects the radiation patterns. An unsymmetrical location on the vehicle will produce a skew, as shown in Figures 6.18 through 6.21. These plots were made using 4nec2. The command sequence is F7 (or Main → Calculate → Generate) → select Far Field, Default Patterns, 2° resolution, → Generate button. Each pattern is a cut through the maximum gain point. The patterns are for total gain, combining gains for vertical and horizontal polarization. There are no nulls, except at the ground elevation, because the car produces significant horizontally polarized waves. The negative θ angles on the left side of the elevation plots are to indicate that this portion of the plot is an elevation at 180° around from the ϕ angle for the right-hand plot. The reader is strongly encouraged to run the calculation and examine the plots for the individual polarizations.

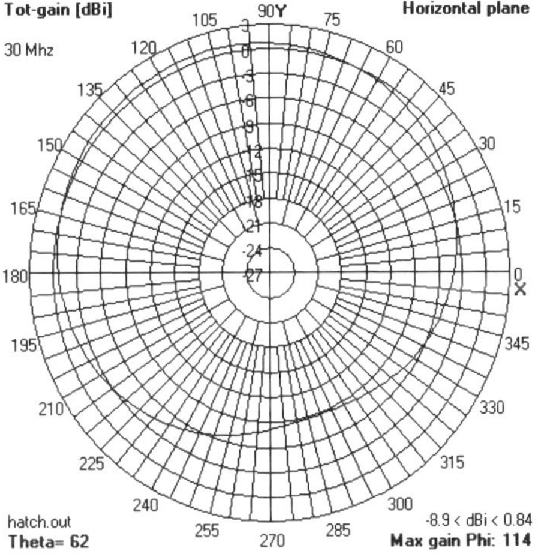

Figure 6.18: An azimuth pattern for the center-loaded antenna of Example 6.7.

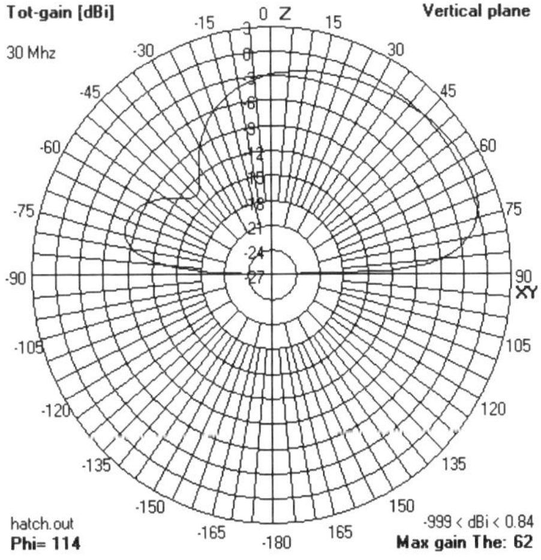

Figure 6.19: An elevation cut at 30 MHz for the center-loaded antenna of Example 6.7.

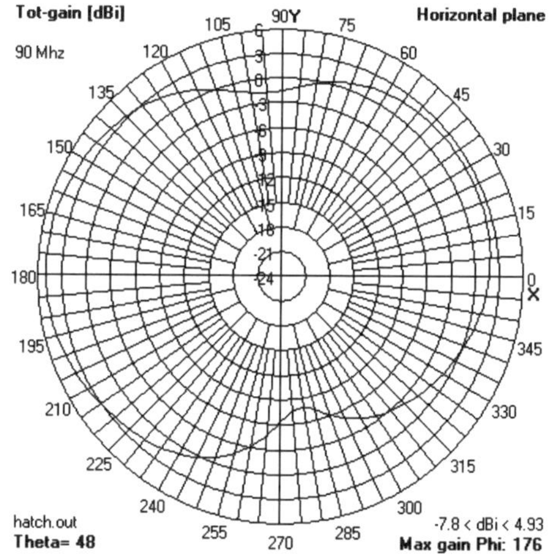

Figure 6.20: An azimuth pattern at 90 MHz for the center-loaded antenna of Example 6.7.

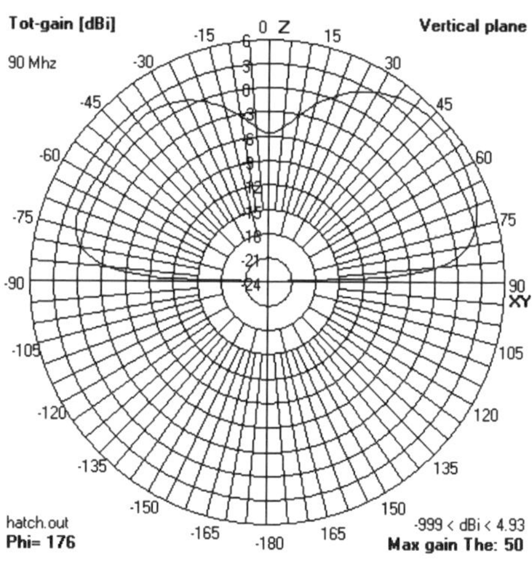

Figure 6.21: An elevation pattern at 90 MHz for the center-loaded antenna of Example 6.7.

Example 6.8 ***HF 80-m Band Whip on the Hatchback***

Many radio amateurs like the challenge of communicating from a car at the long-wave end of the "short-wave" bands. The center frequency of the 80-m band is 3.75 MHz, and [9] has a good deal of information about using a whip on a vehicle at this frequency. He has both numerical and experimental data for a whip which was 110" long, in two equal-length sections. This is 2.794 m, and the diameter of the lower section was 15.9 mm. I decided to put this antenna on the back corner of my hatchback model, tilted back 30° from the vertical to get more separation from the car, and near-vertical-incidence-skywave (NVIS) operating capability. I modified hatch.cpp to make the roof segments align with the side panel segments and saved the program as hatch2.cpp. However, I didn't change the antenna modeling, so I edited the .nec file to change the first GW command. With a 30° tilt, the x and z components of the antenna's length are –1.397 m and 2.42 m. I used 16 segments on the antenna to make the segment length close to that of the car grid. I rounded the radius to 8 mm. A drawing of the result is shown in Figure 6.22.

Figure 6.22: Hatchback with an HF center-loaded whip.

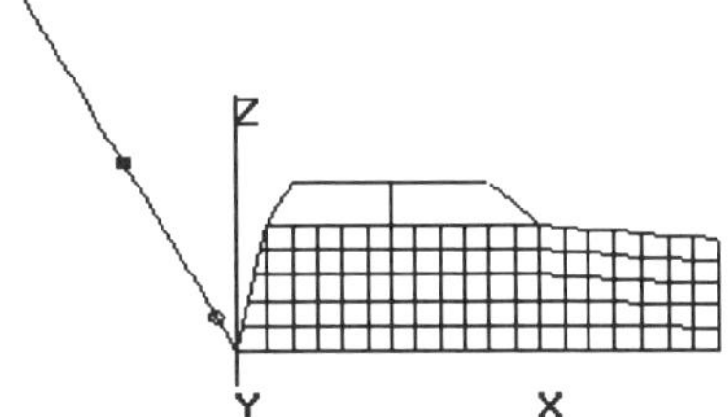

Again, I checked the model in free space and over a perfect ground for validity. The source is in segment 2. In free space $Z_{in} = 0.6 - j1550\Omega$, APG = 0.961. Over perfect ground, $Z_{in} = 0.2 - j1492\Omega$ and APG = 2.067. Both of the APG values are within 5% of target, good enough. Note the sharp drop in radiation resistance, probably due to cancellation of the horizontally polarized component by the ground plane. Over average ground $Z_{in} = 6.6 - j1491\Omega$ and APG = 0.0312, a radiation efficiency of 1.56%. This also implies $R_{rad} = 0.1039$. We are starting from a really low point. The results for base tuning and loading in segment 9 are collected in Table 6.13.

Table 6.13: The 3.75 MHz whip.

Tuning point:

	Segment 2		Segment 9	
Q	200	450	200	450
R_{in}	14.1	9.98	25.548	16.629
E	0.472	0.6679	0.3728	0.5718
A	0.007363	0.010418	0.01079	0.0165723
R_{rad}	0.1038	0.10397	0.27566	0.27528
R_{loss}	7.44	3.3144	16.024	7.1205
R_{gnd}	6.55	6.5617	9.2486	9.2329

The net result is that the A values, the radiation efficiencies, are about 50% higher for the center-loading than the base-tuning cases for the same coil Q. For the $Q = 450$ cases, the combined loss and ground resistances go from about 10 for base tuning to 16.4 for center-loading. This is an increase of 1.6 times, while the radiation resistance increases about 2.7 times. For comparison, the 30-MHz, $Q = 450$ cases from the last example have a radiation resistance increase of 1.7 times and a combined loss resistance increase of 1.56 times. For the 3-MHz monopole of Example 6.6, the radiation resistance increased by about 2.5, and the combined loss increased by about 2.3 times. It appears that the only circumstance for which center-loading is worthwhile is when radiation resistance is a tiny fraction of the combined loss resistances. In the present example, base tuning degrades the radiation efficiency from 1.5% to 1% and center-loading brings it back up to 1.6%.

Figure 6.23 shows a plot of the antenna current for the center-loading case. It so happens that the 300W input power is fairly typical of an amateur rig with a power amplifier added to the transceiver. In this setup, the antenna would be radiating about 5W. Note that the shape of the current plot is very similar to that for the Harrison monopole of Example 6.5, with a peak current about 10–15% above the source current.

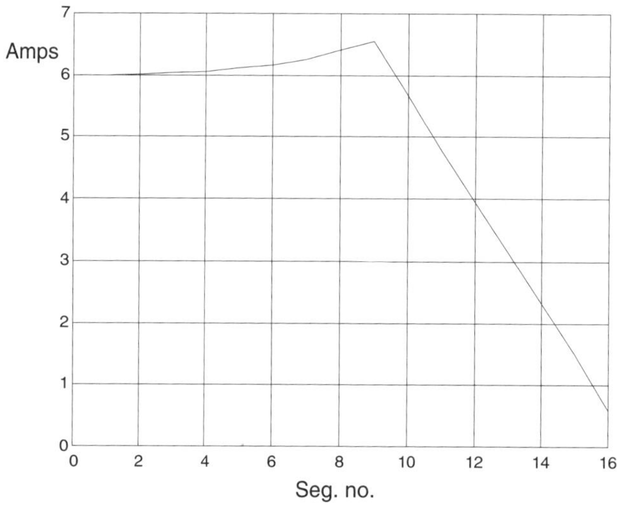

Figure 6.23: Antenna current for the center-loaded whip of Example 6.8. The input voltage is 100V, input power is 300W. Voltage and current are peak values in NEC, rather than rms.

Figures 6.24 and 6.25 show radiation patterns for the center-loaded HF whip. Note the uptilt in the elevation pattern, the desired result of the antenna tilt. Again, the azimuth pattern is a little skewed, and the back gain is about –6 dB down from the front gain.

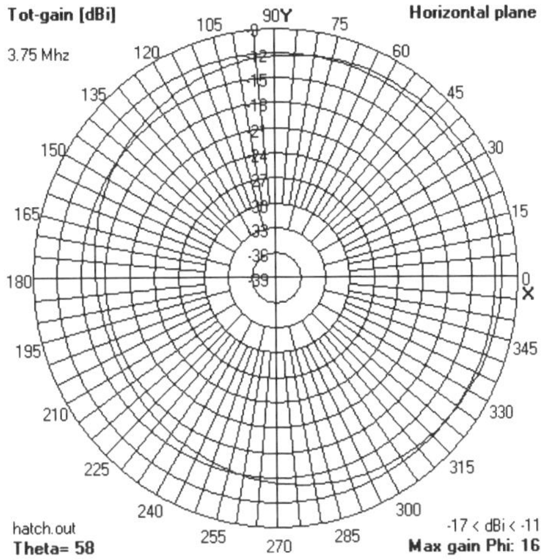

Figure 6.24: An azimuth pattern for the HF center-loaded antenna of Example 6.8.

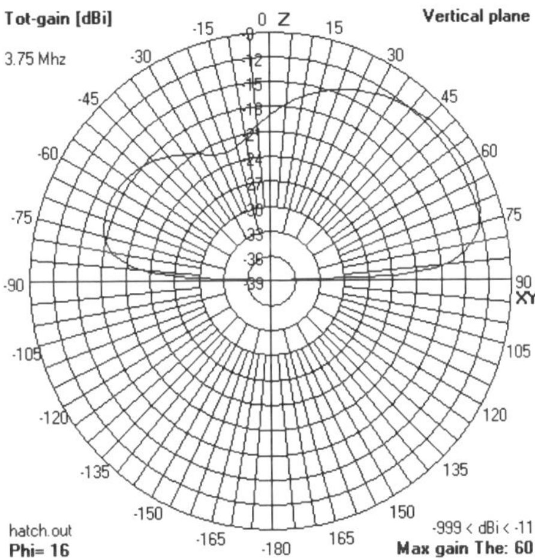

Figure 6.25: An elevation pattern for the HF center-loaded antenna of Example 6.8.

In summary, we see that coil-loading is a good concept, but the Q of the coil must be very high to get a significant improvement in most cases. See coilQ.txt for excerpts from a NEC-LIST discussion on high-Q coils. I chose 200 as a readily achievable value, and 450 as the practical limit anyone claimed.

6.4 Using Resonance

We have presented a few examples of resonant small antennas, most recently Example 6.4 for volume-loading. In that example, the wire's inductance was resonated against the capacitance of the end volumes. It is clear from the top-loading section that an antenna made with wire radials can be made resonant either by using thin enough wire in the radiator, or adding a lumped inductor at the junction between the radiator and the radials. Example 6.2, the inverted-L on a radio box, shows that the impedance can be greatly improved when the radio box is part of the antenna. It seems reasonable that a more effective top-load would achieve resonance in such a case. In this section, I present several examples to explore the practical application of these ideas. Also, a couple of examples using open-ended resonant coils are given.

Example 6.9 **An HF Monopole Study**

Suppose a radio amateur named Hamlet lives in a house on a small city lot, say 70' by 120', and he wants to operate on the 80-m band. He doesn't want to get too far off the ground so let's say that 4 m is his height limit. To begin, Hamlet wants to try using only #14 copper wire because he bought a 500' roll of it for a previous antenna project. He wants to try top loading but doesn't want to put up too many support posts, so the initial study is with 4 radials 5 m long. He knows he'll need ground radials as well, so he'll try 8 radials 5 m long. This will make the antenna occupy about half his lot width. Using tophat.in and tophat.exe, Hamlet constructs a model with 20 segments in the vertical radiator and 25 segments in each radial. The model is set 0.1 m above the x,y (ground) plane, with the source in segment 2. Testing the model in free space and over a perfect ground, with no copper loss, gave APG values of 1.003 and 2.006, very good. The impedance in free space was $1.73 - j334\Omega$, and over perfect ground it was $3.5 - j272\Omega$. Then adding copper loss and average ground parameters he finds $Z_{in} = 11.1 - j272\Omega$, $E = 0.9612$, and $A = 0.12374$. Next Hamlet tries base tuning and tuning with a coil near the top in segment 19. The results are collected in Table 6.14.

Table 6.14: Results for the resonated radial-loaded monopole. Q = 200 in both cases.

	Segment 2	*Segment 19*
R_{in}	12.45	14.8
E	0.8562	0.8565
A	0.11	0.1095
R_{rad}	1.37	1.62
R_{loss}	1.79	2.12
R_{gnd}	9.3	11.1

The radiation efficiency is about the same and is about 10% (1 dB) below the untuned value. Instead of using a lumped tuning coil, how about making the radiator into a coil? Through a trial-and-error process, Hamlet finds that an 8-turn coil with a radius of 0.2734 m, wire length 13.66 m works. The radiation efficiency is still only 11.1%. Doubling the radiator's wire diameter to #8 improves the radiation efficiency 5% (0.5 dB). Okay, how about using the full yard width for the radials? Making them 10 m long raises the radiation efficiency to nearly 20%. Still using #8 wire for the radiating coil, its radius becomes 0.25866 m, and its wire length is 13 m. Good luck trying to make a radius to five figures! But the inaccuracy can be absorbed in the matching network—Hamlet only has to get close. A drawing of this antenna is shown in Figure 6.26, and the analysis of its impedance components is given in Table 6.15.

Figure 6.26: A 3.75-MHz resonant monopole using radials top and bottom and a coiled radiator. The top radials are 5 m long, the bottom ones are 10 m long, and the antenna height is 4 m.

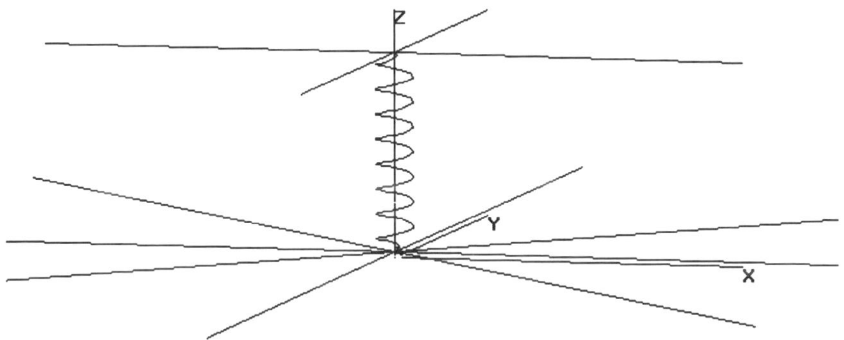

Table 6.15: Performance and parameters of the 3.75-MHz monopole shown in Figure 6.26. L_{equiv} and C_{equiv} are the equivalent series-circuit elements derived by calculating the antenna's input impedance at two frequencies, either side of resonance. See datanec.m. $Q_{coil} = \omega L_{equiv}/R_{loss}$.

R_{in}	7.14
E	0.902
A	0.1967
R_{rad}	1.4
R_{loss}	0.7
R_{gnd}	5.04
Q_{ant}	52.47
L_{equiv}, μH	15.9
C_{equiv}, pF	113.2
Q_{coil}	535.7

A Q of 50 corresponds to a matched half-power bandwidth of only 4%. From eq. (3.19) and (3.21), an SWR = 2 bandwidth is $1/(\sqrt{2}Q)$. The proposed antenna would have a $BW_2 = 0.0135$, or 50.5 kHz at 3.75 MHz. This could mean frequent retuning if Hamlet moves around the band (3.5–4 MHz) a lot. His friend Guglielmo suggests that if Hamlet is willing to put up support posts for those four radials, why not try a framed volume for loading. The increased capacitance should give a lower Q and better bandwidth in the same space. Going back to all #14 wire, Hamlet found that the structure in Figure 6.27 is resonant at just over 3.7 MHz. Its performance is similar to the top-radials-and-coil design, except with nearly double the bandwidth. Its performance analysis is given in Table 6.16.

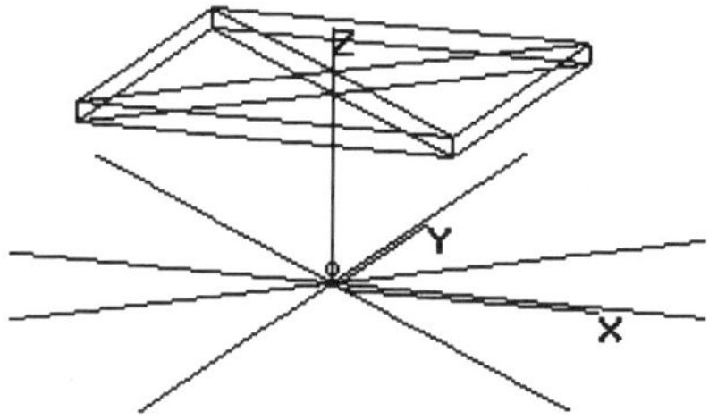

Figure 6.27: A 3.75-MHz nearly-resonant monopole using a volume top load, bottom radials, and a straight radiator. The top load is 7.07 m square, or 5 m from center to corner. The radiator is 3.6 m tall, giving an overall height of 4 m. The bottom radials are 10 m long, not all shown. While the *z* axis goes through the volume, the radiator stops at the volume's bottom so that the current goes out the bottom radials, up the verticals at the corners, and back in along the top radials.

Table 6.16: Performance analysis for the volume-loaded monopole of Figure 6.27.

Z_{in}	$7.1 + j4.2$
E	0.9225
A	0.196
R_{rad}	1.393
R_{loss}	0.55
R_{gnd}	5.16
Q_{ant}	28.03
L_{equiv},μH	8.45
C_{equiv},pF	218
Q_{wire}	362
$BW_{2,}$kHz	94.6

The loss resistance is less because there is less wire in the high-current part of the antenna, but the inductance of the straight-wire radiator is an even smaller fraction of the coil's inductance, so the equivalent coil Q is smaller.

So, what is Hamlet going to do? He's going to think about things, of course. He's learned that the main thing about ground loss is that longer, and probably more, ground radials help a lot. Radiation resistance is pretty well fixed by height. Volume loading gives much more capacitance and bandwidth than radial loading. In many cases, volume loading can be used to achieve resonance without a coil or coiling the radiator. Coiling the radiator is less lossy than using a compact inductor of typical Q. If he does put down more and longer radials in the directions his yard allows, he'll raise the radiation efficiency but decrease his bandwidth.

Example 6.10 **A Resonant Coil Radiator on a Perfect Ground**

An electrically small coil driven at one end and open at the other is generally called a *normal-mode helix* (NMH). It is called normal-mode because the radiation peak is perpendicular (normal) to the axis of the helix. An electrically large helix has its peak radiation along its axis. NMH antennas have been used on personal radios operating in the VHF band for many decades [10]. In the 1990s, many people studied other ways of packing a lot of wire in a small volume to produce resonant antennas, such as meander lines and fractal geometries. S. R. Best [11,12] has done numerical and experimental studies on some of these and concluded that, for a given operating frequency and occupied volume, once enough wire is in there to get resonance the radiation resistance and Q are independent of the wire geometry. The NMH is quite efficient when wire loss is considered and is easy to model, so I have worked up three cases to illustrate its characteristics. The program that generates the wire list is called monocoil.cpp/exe and it reads monocoil.in to generate monocoil.nec. The program uses a function coil(), from meshes.h, that generates a coil starting on the z axis, goes to the specified radius in a half turn, spirals up, and goes back to the z axis in a final half turn. It uses 20 angle steps per turn. The number of turns can be a non-integer, but the smallest change is 0.05 turns because of the 20-step resolution. For these cases, I used 300 MHz as the target resonant frequency to go with the canonic results in sections 6.0 and 6.1. I used 0.8 mm as the wire radius, approximately AWG#14. The resonant frequency is strongly dependent on the wire length and weakly dependent on the wire radius. To get resonance in the specified height and coil radius, I adjusted the number of turns to get as close as possible, then trimmed the coil radius, so the radius numbers aren't quite what they should be. The 2.25-turn helix of Table 6.17 is shown in Figure 6.28. You can see that it doesn't take much wire to get resonance if you have a wide space. The table shows that the wire gets longer as the space narrows and the number of turns has to be increased. A quarter wavelength at 300 MHz is 250 mm and even the shortest entry is longer than this. The standing wave on the wire has the character of a quarter wave, with zero current at the open end and the first maximum at the feedpoint. Since it is in less height than the corresponding free-space wave, this type of antenna is sometimes called a *slow-wave* structure.

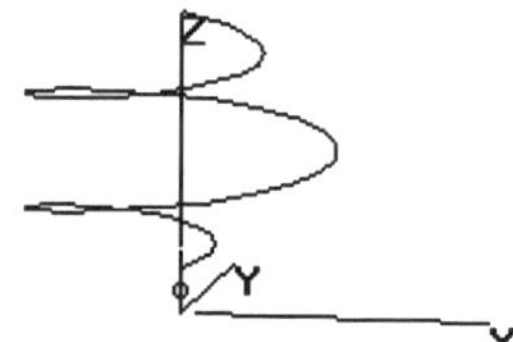

Figure 6.28: A drawing of the NEC model for the second helix in Table 6.17.

Table 6.17: Properties of three small resonant helix antennas. *R* is the turn radius, *L* is the total wire length.

R, mm	f_{res}, Mhz	R_{in}, Ω	*Q*	Turns	*L*, mm
12.55	300.1	2.23	148	4.8	350
24.9	300.4	2.08	120	2.25	294
48.4	299.8	2.11	99	1.25	265

Comparing the helix Q values with the Chu and Wheeler values of Table 6.1, and the thick wire values of Table 6.4, it can be seen that the helix has better Q for its size as it is made thinner. That is, its Q does not rise as fast as the other cases. In [12], Best points out that these single-wire monopoles don't perform the same as other equal-sized monopoles, of which he gives examples. Looking back at Examples 6.3 and 6.4, you can find equivalent top-loaded antennas with much better Q values. The radiation resistance for a uniform-current monopole of height 0.05λ is about 4Ω, almost twice the helix values. The ratio of average to input current must be about 0.7, and the high L to C ratio, are the major reasons for the higher Q of the helix compared to top-load designs.

Example 6.11 **A Cordless Phone Antenna**

In 1996, I designed a 900-MHz band cordless phone antenna set [5] as a demonstration of design possibilities for volume-loading. The antennas were fabricated and installed by technicians at South Dakota State University's College of Engineering, where I was a professor. I had a cordless phone that used channels around 903 and 927 MHz, so I designed the handset and base unit antennas for the mean, 915 MHz. This example illustrates a similar design problem: a handset radio box is to have a volume-loaded antenna entirely within the handset's plastic case. The wavelength is 328 mm, so the handset radio box at 90 mm is just over a quarter wave high. The loading volume will be a plate that parallels the top of the radio box and has the same area. The radio box is 90 × 40 × 14 mm, and the allowed height above the box is 1", 25.4 mm. I used the program VLSD1.cpp/exe to generate candidate designs. The wire-grid model of the final design is shown in Figure 6.29 and the antenna description for the program is given in Listing 6.4 below.

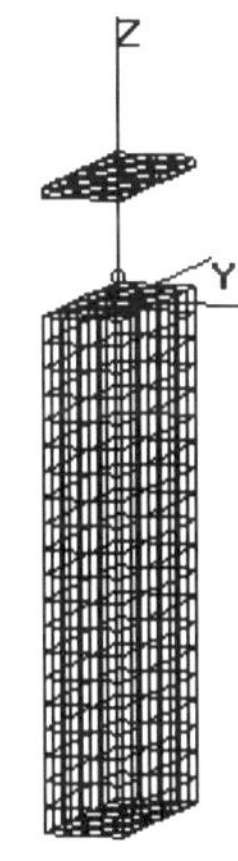

Figure 6.29: Wire-grid model for a volume-loaded antenna on a radio box. The box is 14 × 40 × 90 mm. The antenna's loading plate is 14 × 40 × 1.6 mm. The design frequency is 915 MHz.

Listing 6.4 VLSD1.in

```
2 7 2
0.202 915 0
4 8 1
-7 7 -7 -7
-20 -20 20 -20
21 21 21 22.6
4 8 20
-7 7 -7 -7
-20 -20 20 -20
-90 -90 -90 0
```

```
This file gives the geometry parameters for up to 3 boxes
and a wire joining two of them. It is the input file from
which vlsd1.exe generates the NEC wire list, in.nec. All
dimensions and coordinates are in mm. The data is as fol-
lows, by lines:

1. No. of boxes, no. segs in the antenna wire,
   segment no. for the source.
2. Antenna wire radius, frequency in MHz, conductivity.
   The remaining lines are in blocks of 4 for each box:
   1. No. of wires on each edge.
   2. X coordinates of four corners that define the box.
   3. Y corner coordinates.
   4. Z corner coordinates.
```

The program allows one to put the box anywhere in space, but it was convenient to put the base of the antenna wire, the top center of the radio box, at the coordinate origin. That way, I only had to change the plate's z values to adjust the wire length. A #26 wire about 20–22 mm in length gets the input resistance in the 50Ω neighborhood with small reactance. The reactance can be tuned with either the wire length, which also changes the resistance, or the plate thickness. I settled on a wire length of 21 mm and a plate thickness of 1.6 mm. The final results for the no-loss case are given in Table 6.18.

Table 6.18: Handset model calculations.

MHz	Z_{in}, Ω	*APG*
903	$47.2 - j2.57$	0.976
915	$49.2 - j0.003$	0.976
927	$51.3 + j2.32$	0.976

When copper loss is added, the efficiency is 99.6%.

6.5 Summary

The material in this chapter is a mix of physical design ideas and good modeling practice. We examined ways to get better bandwidth and lower matching loss with our small antennas. Of the methods presented, volume-loading is the most successful technique by far. It should be used whenever it's mechanically feasible. If the space allowed is so small compared to wavelength that resonance can't be achieved with volume-loading and a reasonable wire size, the next best thing to do is coil the radiating wire to get enough added inductance. Lumped coils of typical Q just have too much loss resistance.

We have made a start on examining antennas in their working environments. Antennas over the earth operating in the HF band have greatly reduced radiation efficiencies because of the waves penetrating the ground. On the other hand, VHF and UHF antennas can be helped greatly by the radio box on which they are mounted.

The modeling part of the design process has been illustrated many times. Whether a very simple model is written directly as a .nec file, or extensive code is written to translate a physical description into a wire list, the model must be examined and tested for validity. 4nec2 includes a geometry validation function, which is sometimes too conservative, that should be run. You should examine the drawing of the structure carefully from several perspectives to see that things are where you want them. Then run the model in free space and/or over a perfect ground with no conductor loss to test the average power gain. Then you are ready to test the model variation with segment size. Input impedance and input power are based on a single number, the current in the source segment. On the other hand, structure loss and radiation gain are calculated by summing contributions from all the currents, so they should be more accurate and less changeable with segment size. In some cases, the EX 5 source's dependence on segment length/radius can be used to tune a model that otherwise won't converge or give a good APG without becoming very large in segment number. Even with processors running over a GHz and gigabytes of RAM, a problem can grow to long run times.

Modeling is now everybody's favorite part of the design cycle. In the end, we want a real physical object, so building and testing has to be done as well. The hope is that numerical modeling will weed out bad choices and get the design in the right neighborhood.

References

[1] R. C. Johnson, ed., *Antenna Engineering Handbook*, 3rd ed., McGraw-Hill, 1993.

[2] R. W. P. King, *Tables of Antenna Characteristics*, IFI/Plenum Press, 1971.

[3] N. N. Rao, *Elements of Engineering Electromagnetics*, 3rd ed., Prentice-Hall, 1991.

[4] Wen Geyi, "A Method for the Evaluation of Small Antenna Q," IEEE Trans. on Antennas and Propagation, vol. 51, no. 8, pp. 2124–9, August, 2003.

[5] D. B. Miron, "Volume Loading—A New Principle for Small Antennas," ACES Journal, vol. 14, no. 2, July 1999.

[6] U.S. patent 5,986,610, "Volume-Loaded Short Dipole Antenna," issued 16 November 1999.

[7] J. S. Belrose, "Short Antennas for Mobile Operation," QST , pp. 30–35, p. 108, September, 1953.

[8] C. W. Harrison Jr., "Monopole with Inductive Loading," IEEE Trans. On Antennas and Propagation, pp. 394–400, July, 1963.

[9] J. S. Belrose, "Short Coil-Loaded HF Mobile Antennas: An Update and Calculated Radiation Patterns," in *Vertical Antenna Classics*, R. Schetgen ed., pp. 91–98, ARRL, 1st ed. 5th printing, 2001.

[10] J. B. Andersen and F. Hansen, "Antennas for VHF/UHF Personal Radios: A Theoretical and Experimental Study of Characteristics and Performance," IEEE Trans. on Vehicular Tech., vol. 26, no. 4, pp. 349–357, November 1977.

[11] S. R. Best, "A Discussion on the Quality Factor of Impedance Matched Electrically Small Wire Antennas," IEEE Trans. on Antennas and Propagation, vol. 53, no. 1, pp. 502–8, January 2005.

[12] S. R. Best, "A Discussion on the Predictions of Electrically Small Self-Resonant Wire Antennas," IEEE Antennas and Propagation Magazine, vol. 46, no. 6, pp. 9–22, December 2004.

Chapter 6 Problems

Section 6.1.1

6.1 Write code to generate the wire list for an open-ended EDM. Calculate Z_{in} and Q for the cases in Table 6.4. Comment on your results compared to those in the table.

6.2 Write code to generate the wire list for a hemispherically capped EDM. Calculate Z_{in} and Q for the cases in Table 6.4. Note that the overall height includes the cap. Comment on your results with respect to those in the table.

Section 6.2.1

6.3 Write a model for an inverted-L on a PEC plane using 4nec2's SY command to specify the lengths and segment numbers for the two wires. For $h_m = 50$ mm, $R = 100$ mm, plot the input impedance vs. no. of segments for 50 MHz and 300 MHz. Comment.

6.4 For the last case in Table 6.7, find and plot $|X_{in}|$ against wire radius for $0.05 < r_w < 2.5$ mm.

Section 6.2.2

6.5 Derive eq. (6.24), the inductive reactance component for the vertical section of the monopole.

6.6 Model a top-loaded antenna on the radio box of Example 6.2. Again, the operating frequency is 915 MHz and the vertical height is 10 mm. Mount the vertical in the center of the box top and run four radials out to the corresponding points above the box-top corners. Find Z_{in} and Q.

6.7 Use disc() to model a disc-loaded monopole on the box of Example 6.2. It should be 10 mm high by 20 mm radius. Find Z_{in} and Q.

Section 6.2.3

6.8 Example 6.4 considered a volume-loaded dipole in free space. Move that design to a perfect ground plane and find the radiator wire size to get resonance at 3.75 MHz. The antenna is still 4 m high by 4 m diameter, with a loading-body thickness of 0.4 m. Find Z_{in}, Q, and the SWR = 2 matched bandwidth. You can use the hfocta.* files.

6.9 Write a model-generating program using cylinder() to model a cylinder-loaded version of the antenna in problem 6.8. Find Z_{in}, Q, and the SWR = 2 matched bandwidth.

Section 6.3

6.10 From the current plots, it seems that the current in the center segment of a coil-loaded whip is about 15% over the base current. Model the whip's current as two straight-line sections, one going from I_o to $1.15I_o$ at the coil, and then a straight run down to zero at the tip. On this basis, find an expression for the radiation resistance. The main reactance the coil has to resonate is the open-ended half above it. From this, find an expression for the coil inductance.

6.11 Find and plot the antenna current in Example 6.6. Find the ratio of coil current to input current, and the ratio of coil voltage to input voltage.

6.12 Find and plot the antenna current in Example 6.7 for 90 MHz and coil $Q = 450$. Find the ratio of current through the load to input current, and the ratio of load voltage to input voltage. Comment on results vs. expectations.

6.13 For Example 6.7, re-run the radiation patterns, showing the individual pattern for each polarization. Comment.

6.14 For Example 6.8, re-run the radiation patterns, showing the individual pattern for each polarization. Comment.

6.15 Rework Example 6.8 for an antenna 6 m long. This length would be a problem with low bridges and power lines, but the antenna could be spring-mounted to bend, and insulated. An obstruction-detecting retractor system is also a possibility.

Section 6.4

6.16 From Example 6.9, suppose Hamlet is willing to move the support posts toward the edge of his yard. Keeping the volume square as in Figure 6.27, 60' on a side, find the performance parameters as in Table 6.16.

6.17 From Example 6.9, suppose Hamlet is allowed by Ophelia to use the back half of the lot for his antenna, but still with only the four corner supports. This gives him a space 70' by 60' and a rectangular volume. Write a model that extends the radials to the half-lot edges instead of being in a circle. Find the performance parameters as in Table 6.16.

6.18 For Example 6.10, what polarizations would you expect from a normal-mode helix dipole in free space? From an NMH monopole on a perfect ground?

6.19 Model an NMH dipole in free space and plot the radiation patterns for each polarization. Do the same for an NMH monopole on perfect ground. Comment on the differences.

6.20 Model and design an NMH on a radio box to meet the space and performance requirements of Example 6.11. Run radiation patterns for all polarizations. Comment on the results.

CHAPTER 7

Loops and Other Closed-Wire Antennas

7.0 Introduction

In Chapters 2 and 3 we examined the single-turn small loop, one of the two fundamental antenna types. It is a kind of dual to the small dipole. The loop has a conducting path for current back to the source, whereas the dipole's current is reflected back from an open end. When both antennas are vertical they have vertical polarization, but the loop has a figure-8 azimuth pattern and a constant-level elevation pattern, while the dipole has a constant-level azimuth pattern and a figure-8 elevation pattern, all considered in free space. The loop is inductive rather than capacitive, and its radiation resistance depends on its area in square wavelengths, rather than its length. All of these differences make the loop more useful than the dipole in some applications, mostly receiving.

In this chapter we consider the single-turn loop, the solenoid or uniform helix (which may be thought of as several loops in series or an N-turn loop), the contrawound toroidal helix (CTHA), and an example of a folded antenna. We study them mainly in transmit mode, and defer coils wound on ferrite cores to the chapter on receiving antennas.

Like the dipole, the small loop has a high reactance that can be reduced by designing it to occupy more of the available space. We present modeling methods and results for a doughnut, a sheet barrel, and a thick-walled barrel. Going in the opposite direction, the solenoid clearly has more inductance and should have more radiation resistance than the single-turn loop. Does it have lower Q than either the thin or thick loops? Does it resonate in a small size as the open-wire helix studied in the last chapter? The toroidal helix antennas are operated at resonant frequencies and have some interesting properties.

7.1 Thick Loops

7.1.1 The Doughnut

A circle has the smallest ratio of perimeter to area, so for a given area or length of wire, a circular loop should have the smallest inductance and loss of all possible loop shapes. For this reason most loops have been historically made as circles, and the circular loop has received the most study. This does not mean it's the best shape for every electrically small loop application. If the space available for the antenna is a rectangle, for instance, it will be better to fill the space with a rectangular loop than to use a circle that just fits the narrow dimension of the space. Such issues will be considered later; in this subsection we will focus on the circle form. From the formulas for inductance given in Chapter 2, it is clear that making the wire thicker reduces the inductance and therefore the Q. Also, the thicker wire should spread the current out and reduce the loss resistance.

Figure 7.1 illustrates the three dimensional parameters that are important to us in this discussion, the minor radius a, major radius b, and the outside diameter D. The feed-gap angle is present in the wire-grid models but not in the single-wire models. Since we are always interested in what can be done in a given space, we set D and vary a/b. From the figure, $D = 2(a + b)$, so if we call the ratio $r = a/b$, $b = 0.5D/(1 + r)$ and $a = rb = 0.5rD/(1 + r)$. For $r << 1$, b is close to $D/2$, but as r grows beyond 0.1, b starts to shrink significantly. On the basis of thin-wire theory, this should cause the effective area to shrink, reducing the radiation resistance. Thin-wire theory doesn't take into account the possible variation of current around the minor radius. One can imagine that more current would flow on the shorter path (radius $b - a$) inside a thick loop than around the longer path (radius $b + a$) along the outside of the loop.

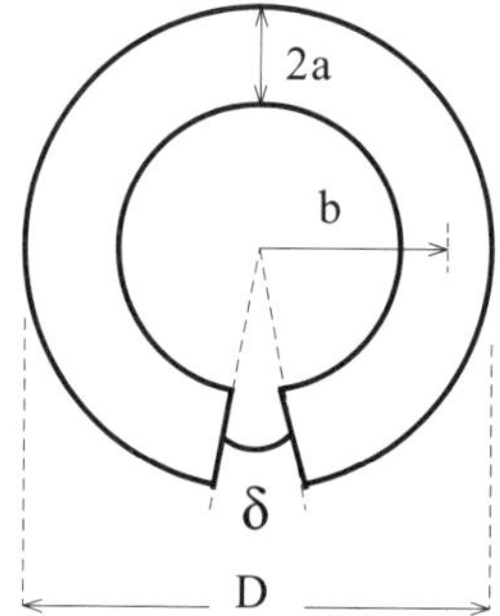

Figure 7.1: Cross-section and dimension labels for a circular loop. δ is the feed-gap angle.

Some results from the simple analysis formulas of Chapter 2, and a modified version of the single-wire loop model given in Chapter 4, are presented in Table 7.1. The equivalent L and C values are for a parallel circuit as described in Chapter 2.

Table 7.1: Performance results for thin-wire loops. Free space, no loss.

$D = 0.1\lambda$.

Simple theory.

a/b	Z_{in}, Ω	*Q*	*L*, μH	*C*, pF
0.01	5.18 + *j*920	297	2.914	3.89
0.05	4.02 + *j*553	219	1.84	5.7
0.1	3.01 + *j*388	195	1.36	7.03
Single-wire model, EX 0 and 32 segments.				
0.01	6.66 + *j*882	218	2.843	3.88
0.05	5.85 + *j*574	166	1.804	6.37
Single-wire model, EX 0 16 segments.				
0.05	4.76 + *j*526	172	1.793	5.62
0.1	3.78 + *j*388	158	1.332	7.45
Single-wire model, EX 5, 175, 35, 17 segments, respectively.				
0.01	9.52 + *j*1052	215	2.869	4.77
0.05	11.5 + *j*806	165	1.817	8.905
0.1	9.62 + *j*631	157	1.402	11.677

The number of segments that can be used in a NEC model is limited by the wire radius, $2\pi b/N_{segs} > 2a$, or $N_{segs} < \pi b/a$. This is why I changed to 16 segments for the EX 0 source at $a/b = 0.1$. From section 4.6.2 and Table 4.5 we see that the reactance gets higher and the APG gets closer to 1 as the number of segments increases with the EX 0 source. On the other hand, I was able to tune the EX 5 source model with the segment number to get APG = 1.001 or better for the two thinner cases, but had to settle for 0.962 for $a/b = 0.1$. From Chapter 2, we know that the impedances we are calculating are on the slope of a resonance curve, so a large part of the variation for a given thickness is due to a shift in the predicted first resonance frequency. This effect is explored after the presentation of the wire-grid model.

Figure 7.2 Wire-grid model for a loop with major radius = 0.3333 m, minor radius = 0.1667 m. The feed gap is 10.445°. The number of steps around the minor radius is 18, and around the major radius it is 35.

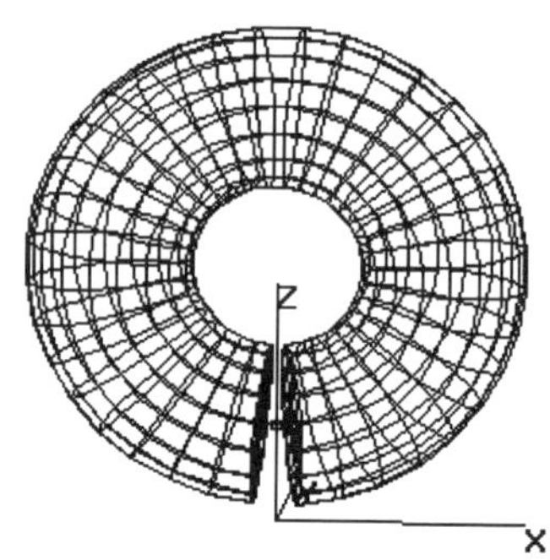

The program group for the wire-grid models illustrated in Figure 7.2 is donut.*. The gridding is approximately square on the side of the loop, and a modified version of symdisc2(), disc3(), is used to grid the flat ends on either side of the feed wire. disc3() puts chord wires around the rings to simulate angle-directed currents. As in the similar case in Chapter 6, I couldn't get the EX 0 source to give a good APG, so I used the EX 5 source and tuned it with the number of segments in the source wire and the gap angle. The gap angle determines the source wire length, so fine adjustment in the segment length/radius can be made with this parameter. The case in Figure 7.2 is for $a/b = 0.5$, so the number of steps around the major radius is twice that around the minor radius. For thinner loops, this ratio grows correspondingly larger, leading to rapidly increasing computation times. For $a/b = 0.05$ and 0.1, I used 14 steps around the disc edge to get reasonable run times. The performance results are collected in Table 7.2.

Table 7.2: Results for the wire-grid thick loop models. Free space, no loss. D = 0.1 λ.

a/b	δ, deg.	Z_{in}, Ω	Q	L, μH	C, pF
0.05	2.6	$8.43 + j690$	166	1.811	7.86
0.1	4.292	$6.84 + j520$	157	1.335	10.88
0.2	8.58	$3.05 + j292$	171	0.866	14.3
0.3	12.88	$1.29 + j175$	207	0.606	16.1
0.5	10.445	$0.294 + j75.6$	354	0.292	26.3
0.7	12.016	$0.0412 + j29.6$	827	0.136	27

The impedance drops dramatically as a/b is increased, and the Q rises sharply after $a/b = 0.1$. Both of these effects are indications that the effective size of the loop is going down as it's made thicker. I displayed the currents for the $a/b = 0.5$ case, and found that the current going around the inside path is 20 times that going around the outside path. Given this fact, it is not surprising that there is an a/b

for minimum Q. I have not tried to nail it down because I believe we have established that the doughnut is not the best loop shape, but I'm guessing it's close to $a/b = 0.1$ and $Q = 150$. Looking back at Table 7.1, the thin-wire models are rather too optimistic.

I thought it would be instructive to see where the various models would predict the first resonance frequency. The expected thin-wire resonance is for $2\pi b/\lambda_{res} = 0.5$ or $f_{res} = 75/(\pi b)$. For $b = 0.5$ m, $f_{res} = 47.7$ MHz. Table 7.3 collects some results for $D = 1$ m.

Table 7.3: Resonant frequencies for $D = 1$ m.

Frequencies are given to the nearest 0.5 MHz.
Single-wire models.
EXn is the EX source type, N_{segs} is the number of segments in the loop.

a/b = 0.01

EXn, N_{segs}	0, 32	0, 64	0, 175	5, 175
f, MHz	46.5	45.5	44.5	42.5

a/b = 0.05

EXn, N_{segs}	0, 32	0, 64	5, 35
f, MHz	46	43.5	39.5

a/b = 0.1

EXn, N_{segs}	0, 32	5, 17
f, MHz	45	39

Wire-grid models

a/b	0.1	0.5
f, MHz	41.5	56

For a given single-wire loop model thickness and the EX 0 source, the resonant frequency decreases as segment density increases. It might reasonably be concluded that the tuned EX 5 source produces the value that the EX 0 source would reach if the segment length didn't get too short. Taking the data altogether, the resonant frequency has a minimum value at some thickness, probably the same thickness as for the minimum Q. The value from Table 6.1 to which we should compare the loop is $Q_c = 35.4$. As practical antennas go, a Q of 157 is pretty good.

Example 7.1 **An Apartment Loop**

Over the years, a number of transmitting loops have been sold for the Hf band. The main advantage is, of course, that a 1-m diameter loop doesn't take much space. It can be mounted on a tripod or even hung against a wall, using a sturdy anchor. We consider here a 1-m diameter loop with $a/b = 0.1$ and made of aluminum. This choice makes $b = 0.4545$ m and $a = 0.0455$ m. The tubing diameter is $2a = 91$ mm = 3.58 in. I chose to assume the owner is on the second floor of a wood-frame apartment building (so I wouldn't have to consider the effect of the steel framing in some buildings) and the ground is poor and dry soil, $\sigma = 1$ mS/m, $\epsilon_r = 5$. I chose to analyze the loop at frequencies near the band centers for the amateur bands from 6 m to 40 m, skipping the narrow ones. As in the studies above, the calculations were done at two frequencies separated by 0.2 MHz. The APG and efficiency numbers were taken from the higher frequency result, since they vary little over that range. I thought it would be interesting to see what the applied voltage at the loop is for an input power of 100W, a typical transceiver power level. The NEC data was processed through the MATLAB functions datanec.m, antres.m, and antvolt.m. The results are collected in Table 7.4.

Table 7.4: Results for a 1-m diameter aluminum loop in the amateur radio bands from 6 m to 40 m. bw2 is the bandwidth in kHz between SWR = 2 points.

MHz	Z_{in}	*bw2, kHz*	*A*	*E*	R_{rad}	R_{loss}	R_{gnd}	V_{100} *Vrms*
52	$68.3 - j672$	1310	0.601	0.9986	41.1	96m	27.2	817
28.9	$4.59 + j465$	105	0.442	0.979	2.03	96m	2.47	2169
21.3	$0.695 + j243$	32.2	0.41	0.936	285m	44m	366m	2913
14.2	$0.127 + j136$	8.3	0.347	0.786	44m	27.1m	55.8m	3828
7.2	$0.0225 + j63$	1.76	0.127	0.251	2.86m	16.9m	2.79m	4222

As the operating frequency decreases, the wavelength increases and the loop resistance also decreases. This means that the loop current has to grow to get 100W into it. Even though the reactance is also dropping, it's not as fast as the current grows, hence the increase in required voltage. For 10 m and down, the loop has excessive inductive reactance, so it is impedance-matched by an L-section using two variable capacitors. The one at the loop will see essentially the same voltage as the loop. At 40 m, the bandwidth is essentially one voice channel. I recall at least one commercial loop that used a motor to retune the capacitor to follow the transmitter frequency. Looking at the resistance breakdown is quite interesting. Roughly half of the radiated power goes into the air and the other half goes into the low-conductivity ground. This is almost like radiating into free space. It would be interesting to run the simulations for higher-conductivity grounds, as requested in the problems.

7.1.2 The Barrel Loop

The optimum doughnut loop doesn't take up too much of the space available, in terms of a cylinder 0.1λ high and wide. It has been long known that loops spaced apart on the same axis, connected in parallel, have lower inductance than just one loop. Taking this to its logical limit, we can imagine a metal sheet rolled into a tube shape, but leaving a slot so it can be driven like a loop. Figure 7.3 illustrates a wire-grid model of this idea.

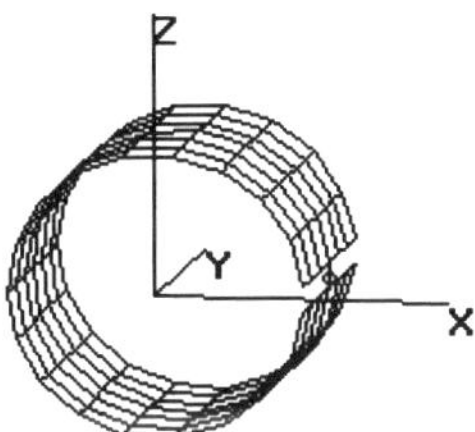

Figure 7.3: Wire-grid model of a thin-walled barrel loop. It is 1 m wide and 1 m long. The arc steps are 20° for clarity, and the feedgap angle is 10°. The grid cells are approximately square, even though the perspective view makes them appear otherwise.

Unlike the doughnut, the barrel has only three radius values, one for the arc segments, one for the *y*-directed segments, and one for the source wire. This fact makes it fairly easy to write directly as a 4nec2 file, as shown in Listing 7.1.

Listing 7.1: thinbarrel.nec

```
CM Loop cylinder, thin wall.
CE
SY N=58 'No. angle steps.
SY ga=5 'Half gapangle, deg.
SY ks=0.7
SY b=0.5 'Barrel radius.
SY da=(360-2*ga)/N 'Arc angle step.
SY al=da*b*pi/180 'al=arc length for loop segment.
SY ang2=360-ga
SY w=1 'Barrel length.
SY nw=int(0.5+w/al)
```

```
SY dw=w/nw 'Loop width step.
SY ysrc=dw*int(0.5+nw/2)
SY rw=al/(2*pi),ra=dw/(2*pi)
SY dz=b*sin(ga),dx=b*cos(ga)
SY rs=ks*ra
SY nls=nw+3
GW 1 3 dx ysrc -dz dx ysrc dz rs
GA 2 N b ga ang2 ra
GM 1 nw 0 0 0 0 dw 0 2
GW nls nw dx 0 dz dx w dz rw
GM 1 N 0 -da 0 0 0 0 nls
GE 0
GE -1
EX 0 1 2 1 100 0
PT -1
FR 0 1 0 0 30 0.01
RP 0 46 46 1002 0 0 2 2
EN
```

The program is very simple in concept. The first wire is the source wire. Then a GA command is used to generate the basic loop almost-circle. A GM command repeats this arc in the *y* direction. Then a wire is drawn along the upper edge of the feed gap, and a GM command is used to repeat it around the *y* axis. All the SY commands are to take the basic dimensional parameters and generate the appropriate start, stop, and segmentation parameters.

Initially I tested the model with a one-segment source wire and the EX 0 source. The source wire radius was made equal to the arc segment radius. This approach gave APG values around 1.03. I then went to three segments on the source wire and the APG values went down to around 1.02. I discovered that the APG value could be tuned with the source wire radius. Making the feed gap angle equal to the arc segment angle meant that the feedgap shrinks as the grid density is increased. This effect caused the impedance to rise as the grid density increased. I then did a run holding the feedgap angle constant at 10°, and this showed a tendency to converge. Table 7.5 shows results for this last combination.

Table 7.5: Impedance vs. grid density. The loop is 1 m by 1 m, *f* = 30 MHz, and the feedgap angle is 10°. The source is EX; the source wire has three segments. *N* is the number of arc segments, and *ks* is a factor that multiplies the arc wire radius to give the source wire radius.

N	Z_{in}	ks
29	$27.3 + j401$	1.2
35	$22.8 + j371$	0.975
49	$19.5 + j349$	0.8
58	$18.8 + j346$	0.7
70	$18 + j341$	0.62

For $N = 70$, the Q is 48, a pretty good value. It is three times better than the optimized doughnut, but the volume-loaded dipole of Example 6.4 is better still. The high resistance values are an indication that we are fairly high up on the resonance curve. For $N = 58$, the resonant frequency is 37.7 MHz.

The preceding is all about a thin-walled barrel. What happens if we make it a little thick? I extended thinbarrel.nec by adding more SY, GA and GM commands to generate another layer of arc and straight lines on the inside for a thickness t. The new file is thickbarrel.nec. For a thickness of 0.05 m, 10% of the outside radius, the model convergence was a little better than before. At $N = 58$, $Z_{in} = 15.3 + j311\Omega$. However, the Q is 54.7, showing that thicker is not better in this case. Keeping all the current on the outside gives the best performance.

Example 7.2 A Strap Loop on a Cell-Phone Radio

This example concerns the 850-MHz cell-phone band. The assumed box is 80 mm high by 40 mm by 20 mm. Since the radio and the antenna are both inside a plastic case, I assumed I could use the space above the radio and wrote a C++ program to generate a wire list for a rectangular loop. The file set is straponbox.*. Since the loop is simulated by grids of straight lines, each line in a grid has the same radius, so I defined one line and used the GM command to generate the rest. The box() function was written to represent a box in an arbitrary orientation, so it was not written with GM commands, although boxes with coordinate-plane orientation could easily be constructed with a lot fewer lines of code using GM commands. Figure 7.4 shows the wire-grid model, using 4-mm segment lengths for the cells. The loop feed gap is also 4 mm. It is driven by a tuned EX 5 source. I tried a 10-mm loop first, but didn't get a good enough impedance, so I enlarged it to 20 mm. In free space, with no loss, the circuit performance results are $Z_{in} = 7.826 + j130\Omega$, $Q = 26.1$, $L = 15.4$ nH, $C = 0.83$ pF. At 850 MHz, $\lambda = 353$ mm, which makes the loop dimension 0.057λ. The low Q shows that the loop is getting a lot of help from the box. Since the input reactance is positive, it can be easily matched with two capacitors.

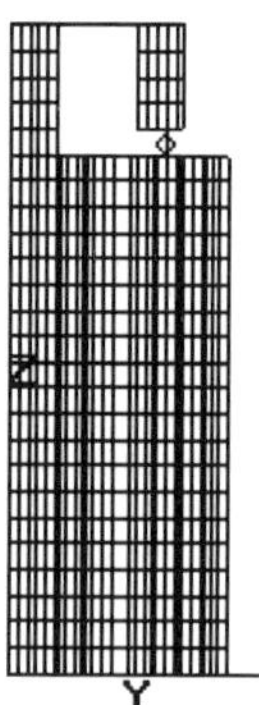

Figure 7.4: Wire-grid model for a strap loop on a cell phone. The narrow face of the radio box is in the *x-z* plane and so is the near edge of the loop. The loop is 20 mm in all three directions, occupying the near half of the box top.

What about that box help? The far-field patterns are shown in Figures 7.5 and 7.6. The elevation pattern shows a dip, characteristic of a vertical dipole, but not very deep. The horizontal pattern has very little variation, again like a vertical dipole. So the box is the major radiator, with the loop partially filling in the high-elevation null.

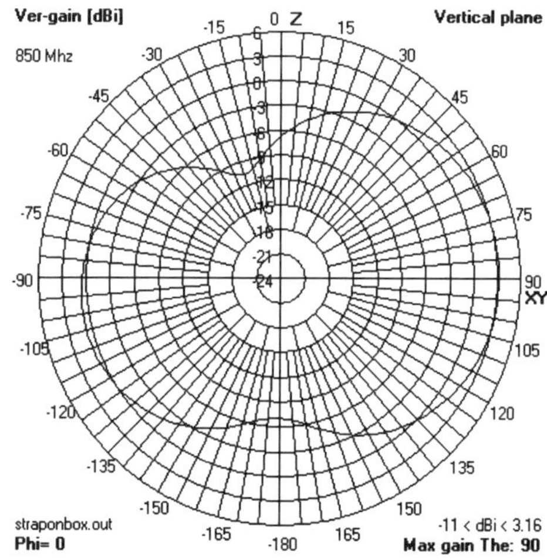

Figure 7.5: Vertical radiation pattern for the cell-phone of Figure 7.4.

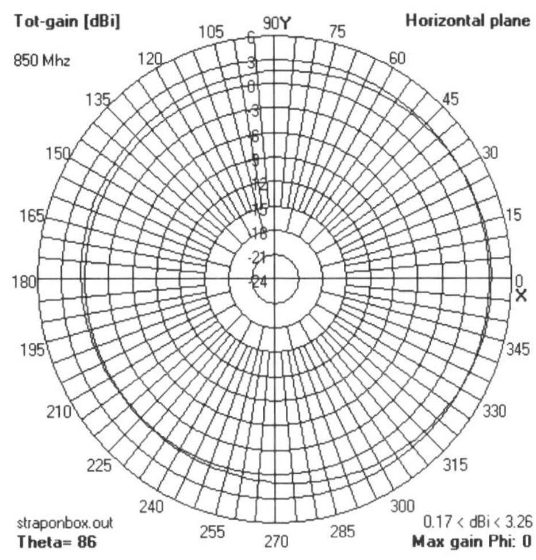

Figure 7.6: Horizontal radiation pattern for the cell-phone radio of Figure 7.4.

Example 7.3 A Long Strap Antenna

A major concern for portable phone designers has been the fact that in urban environments the radios often have to work in standing wave patterns. These are caused by the large and changing amounts of metal in the vicinity of the radio, which can cause electric-field nulls. Proposed solutions usually involve using two antennas to get some kind of diversity advantage, separation in space, different polarization, or different field component responses. In the last example, we have a loop that acts like a dipole because it's dominated by the currents on the radio body. If that antenna and radio are placed near and parallel to a perfect ground, its radiation resistance drops to around 1Ω, and its reception ability falls off. This happens because the dipole mode is shorted by the ground plane and the loop mode isn't strong enough to pick up the slack. Maybe it's because of results like these that I haven't seen a portable phone with any diversity on the market.

Well, okay so the little loop won't do the job—how about putting it on a long face of the radio and covering the entire side? Figure 7.7 illustrates the wire-grid model of this idea. The space between the radio and the strap is 12 mm, the feed

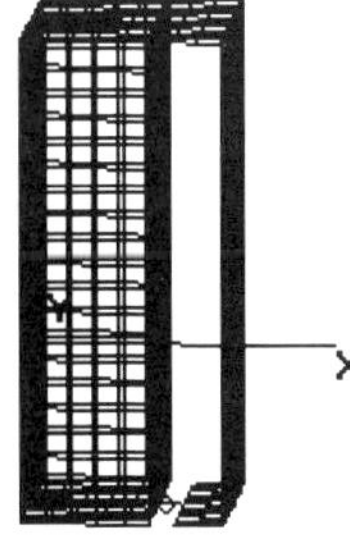

Figure 7.7: Cell radio box with long strap antenna, in vertical position.

gap is 4 mm, and the strap is 40 mm by 80 mm. It's not electrically small any more, but it's low profile.

The strap has a loop area of 960 mm^2, 2.4 times that of Example 7.2. Its perimeter is just above the half-wave resonant length for 850 MHz, and the impedance analysis shows that $Z_{in} = 38.2 - j262.3\Omega$ and $Q = 30$. The parallel LC equivalents are 15 nH and 3 pF. Figures 7.8 and 7.9 show radiation pattern cuts for the radio in free space.

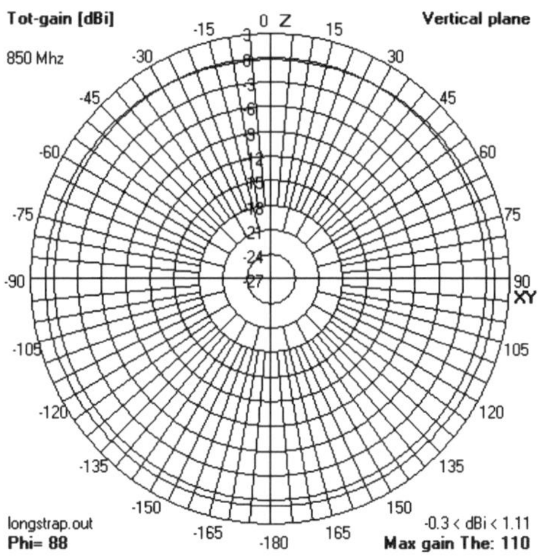

Figure 7.8: Vertical pattern cut for the radio and long strap antenna of Figure 7.7 in free space.

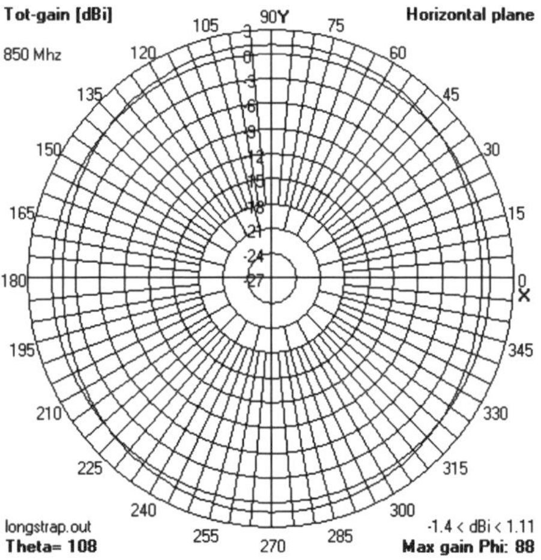

Figure 7.9: Horizontal radiation pattern for the long strap antenna of Figure 7.7 in free space.

It can be seen that there is less than 3 dB variation in the gain in these patterns, showing that neither dipole nor loop modes dominate. The next step is to see how the antenna behaves near a perfect ground. The wire list for Figure 7.7

was developed from that of Figure 7.4 by swapping the height and depth dimensions of the box , which automatically made the strap 80 mm long, and changing the height and width numbers for the loop in straponbox.in. The result is a radio lying on its broad side with the antenna parallel to the *x* axis. To get Figure 7.7, a GM command was used to rotate the model about the *y* axis by 90º. So, to test the model against a ground, I took out the angle increment and put in a *z* increment variable. The command reads

```
GM 0 0 0 0 0 0 0 h 1
```

Figure 7.10 shows an elevation pattern with the radio parallel to and 5 mm above a perfect ground plane.

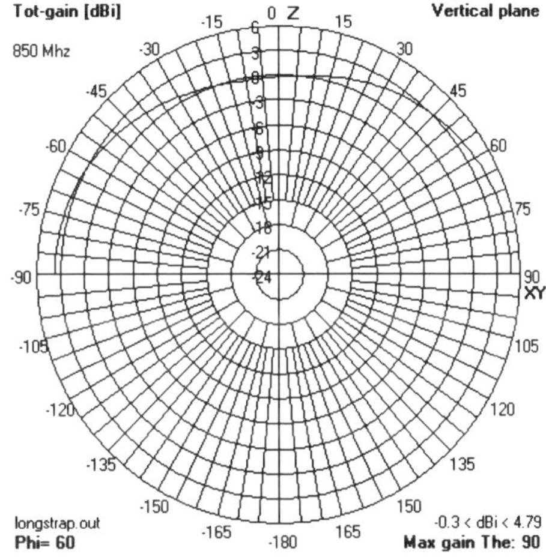

Figure 7.10: An elevation pattern for the cell radio of Figure 7.7 lying 5 mm above a PEC ground.

Again, this pattern shows no serious nulls, although there is a 5-dB spread in the gain. Gain is a ratio of power density in a specific direction to the average power density. It doesn't tell you how hard it might be to actually get a signal out. To get a better idea, I used the Sweep facility of 4nec2 to calculate Z_{in} for $h = 5$ to 180 mm, just over a half-wave distance. The result is plotted in Figure 7.11. Here, instead of being shorted out by the ground plane, the input resistance goes up by less than two over the free-space value. Thus, if the antenna is conjugate-matched in free space, there will be less power (maybe –3 dB) transmitted at the worst position. This is a lot better than the dipole mode.

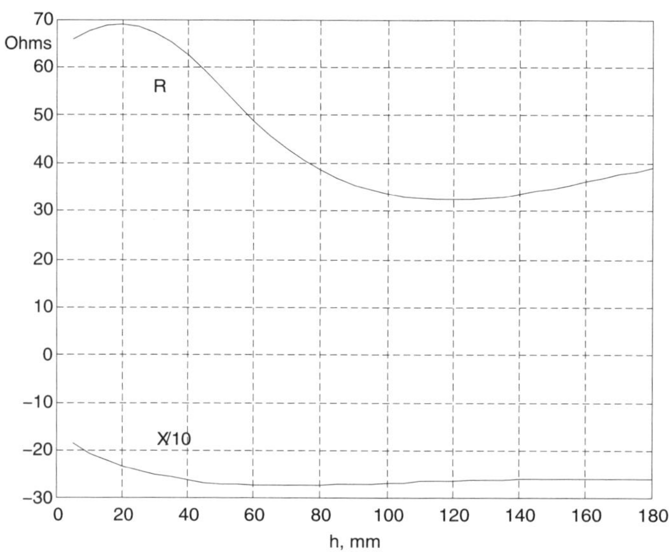

Figure 7.11: Input impedance vs. height above the ground plane for the long strap antenna and cell radio. The radio is parallel to the ground plane.

Well, it transmits. Presumably it will receive also, but under what conditions? NEC provides for excitation by a plane wave, either linearly or elliptically polarized. I used an LD 4 command to conjugate match the antenna for free space, a PT 0 command to display the current in the source wire, and an EX 1 command to apply plane waves from several different directions to test the antenna in free space and over the PEC ground plane.

```
LD 4 1 2 2 32 238
EX 1 2 2 0 0 0 0 90 90
PT 0 1 2 2
```

Wire 1, segment 2 was the voltage source location, so both the LD and PT commands specify this segment. The results are collected in Table 7.6. Figure 7.12 illustrates the wave direction and field vector orientation for the various cases. Of the four principal orientations, only one produces no current. This is an orientation in which the $\overline{E}$ vector is parallel to the broad side of the radio, and the magnetic field does not pass through the loop. In the other cases, either the electric field is applied between the strap and the radio box, or the magnetic field passes through the loop, or both. The presence of the ground plane only seems to enhance the nonzero responses. Clearly, this antenna arrangement would be an effective solution to the standing-wave reception problem.

Table 7.6: Received current as a function of wave direction. The source wire is loaded by 38.2 + j262.3Ω. The E vector is in line with the θ unit vector in all cases.

Free space.

θ, ϕ	*I, mA*
0, 0	0.667
0, 90	0
90, 0	0.905
90, 90	0.946

Near PEC plane.

θ, ϕ	*I, mA*
0, 0	0.649
0, 90	0
90, 0	1.07
90, 90	1.098

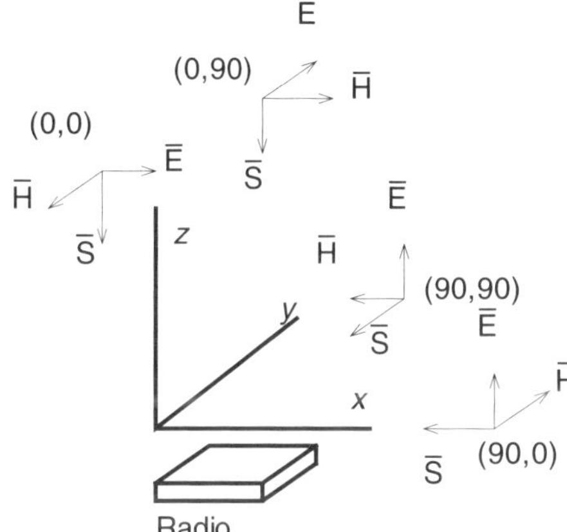

Figure 7.12: Diagram showing wave directions for Table 7.6. The numbers in parentheses are the corresponding (θ,ϕ) values in degrees. S stands for the Poynting vector, which shows the direction of wave travel, and whose amplitude is the wave's power density.

7.2 Solenoid Antennas

In the last section we explored ways of making the single-turn loop fill the space available to get the most bandwidth. Now we go in a different direction, by filling the space with many turns. The equivalent circuit for a small N-turn loop has the same form as that for a single-turn loop, an inductance in series with radiation and loss resistances, shunted by a capacitance. as shown in Figure 2.15. We have analytical help with the radiation resistance and the inductance. In addition to the dimensional parameters in Figure 7.1, let's define the mean diameter as $D_m = 2b$ and the solenoid length as l_c. Since the solenoid is electrically small, we can assume that, in receiving mode, the same magnetic field passes through each turn. Then the total open-circuit voltage is just the series addition of that given by Faraday's law for one turn, equation (3.25), $V_{oc} = -jN\beta AE_{inc}$. Then the effective height defined in (3.26) is just $h_e = N\beta A$. Equation (3.30) relates the effective height to the directivity and radiation resistance. Because the solenoid is small, we may assume its directivity is the same as that for a single turn, 3/2. Solving for the radiation resistance gives us:

$$R_{rad} = 20(NA\beta^2)^2 \tag{7.1}$$

which is N^2 times that for a single turn loop.

Analytical results for the inductance are obtained by assuming the solenoid current can be approximated by a current sheet; that is a way of saying the turns are closely spaced. One simple formula is[1]:

$$L = \frac{N^2b^2}{9b + 10l_c} \qquad \mu H \tag{7.2}$$

However, the dimensions in this formula are inches. I put this in a more general form so that the permeability is explicit and any compatible length unit can be used.

$$L = \frac{3.133\mu_o N^2 b}{0.9 + l_c / b} \tag{7.3}$$

The basic solenoid Q is

$$Q = \frac{\omega L}{R_{rad}} = \frac{5.984}{(\beta b)^3 (0.9 + l_c / b)} \tag{7.4}$$

Clearly, the number of turns doesn't matter in this expression, but when loss is considered, fewer turns gives less wire length and loss. But, remember the sheet-current assumption. The Q goes down with both size parameters. If we make the mean diameter equal to the solenoid length, $Q = 2.06/(\beta b)^3$, and for our canonic 0.1λ cylinder $Q = 66.4$. This is not as good a result as the thin-walled barrel of the last section, and we've probably violated another smallness condition: the total wire length has to be less than 0.1λ for the sheet-current assumption [2]. Essentially, the sheet-current theory was developed by people who were interested in inductor design, not antenna design. The inductor is to be used well below its first resonant frequency and to have as high a Q as possible. The typical inductor has a loss resistance much higher than its radiation resistance. A solenoid used as an antenna should be run at or slightly above its second resonance, where it may have a reasonable radiation resistance and a reactance suitable for the series leg in an impedance-matching circuit.

I've written a wire-list generator in the file family solenoid.* to demonstrate the trends in designing solenoid antennas. Figure 7.13 illustrates a four-turn solenoid in free space. As an antenna, the solenoid has to be driven by some wire arrangement and the length of the wire is significant. Notice that the drive wire and the solenoid form an effective one-turn loop perpendicular to the loops forming the solenoid.

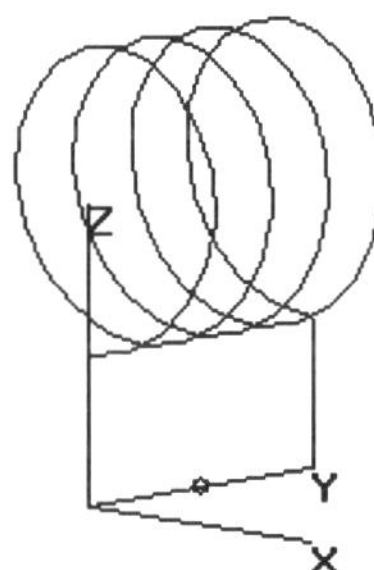

Figure 7.13: A four-turn solenoid antenna. It is 0.1 m long by 0.1 m diameter. The axis is up from the *x-y* plane by 0.1 m, so that the total feed wire length is 0.2 m.

Since we are looking for electrically small designs, I exercised the programs to find resonances at or below 300 MHz ($\lambda = 1$ m) for our 0.1-m cylinder. I looked at both free space and perfect ground environments. The results are collected in Table 7.7.

Table 7.7: Data for solenoids that fit in a 0.1-m cylinder. The solenoid wire has 1-mm radius and no loss. An EX 0 source was used. The source wire was adjusted in segmentation and radius to give the target APG within 1%.
n = number of turns.
l_w = total wire length, mm.
f_1 = first resonant frequency, MHz.
f_2 = second resonant frequency, MHz.
Q_2 = quality factor at f_2.
R_2 = input impedance at f_2.
L and C are the series-equivalent-circuit elements at f_2.

Free space. Diameter = length = 0.1 m.

n	l_w	f_1	f_2	Q_2	R_2 Ω	L, nH	C, pF
1	528	292.07					
2	834	179.13					
3	1144	125.2	278.5	83.9	9.99	958	0.341
4	1455	94.7	229.33	148	5.79	1190	0.404
5	1768	76.03	192.3	279	3.76	1737	0.394

Perfect ground. Axis 0.1 m up.

3	1044	87.99	190.8	26.8	25.25	1125	0.618
4	1355	68.140	155.03	53.9	16.45	1818	0.58
5	1668	55.45	129.79	101	11.4	2824	0.533

Perfect ground. Diameter = axis height = 0.05 m.

3	530	176.03					
4	684	140.71	300.25	40.4	14.39	d616	0.456
5	839	117.02	255.01	67.3	10.32	860	0.45

Free space. Diameter = 0.05 m, length = 0.1 m.

8	1405	111.17	279.51	198	7.45	1680	0.193

The table shows trends, but there are no absolutely simple relationships. Even the second resonance frequencies are not twice the first resonance frequencies. In general, the more wire (more turns) you pack into a given space, the lower the resonant frequencies and the higher the Q. Being over a perfect ground is better than being in free space, even when the solenoid diameter is cut in half. The three-turn entry for the first free-space rows has twice the Q of the half-diameter perfect-ground 4-turn entry, which has the nearest resonant frequency. The last row shows that reducing the solenoid diameter requires more turns to get the same resonant frequency, and the Q gets worse.

The current distribution at any frequency is a standing wave. Figure 7.14 shows the normalized current at the second resonant frequency for the 4-turn free-space solenoid. You can see that the outside turns are carrying slightly less current than the inside turns. For interest, I tried fitting it with a cosine curve. The data is on 80 segments, and the maximum is close to the center. The feed-wire current is not part of this data, and probably completes the cosine. I found that the curve is matched perfectly by $I_c = -\cos(\text{pi}*(\text{segs}-41)/43)$ which converts to

$$I_c = -\cos(\text{pi}*(\text{turns}-2.05)/2.15).$$

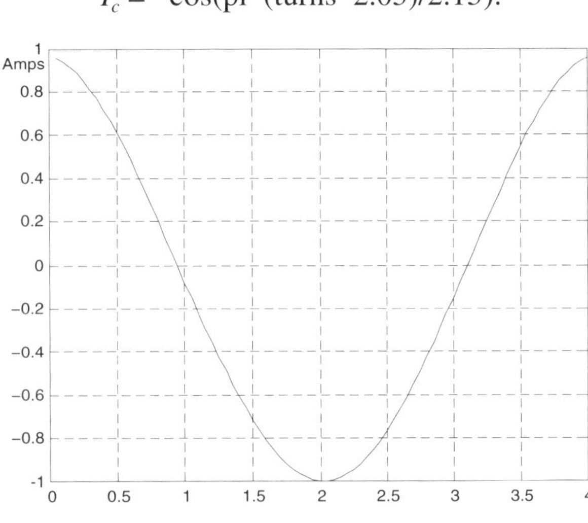

Figure 7.14: Normalized current in the 4-turn solenoid at second resonance.

One could reasonably expect a radiation pattern similar to a one-turn vertical loop, and this is almost the case. For the 4-turn solenoid in free space, the *x-z* plane pattern is within 0.1 dB of constant and the *x-y* plane pattern is the expected figure-8, but with very shallow dips. The latter pattern is shown in Figure 7.15.

When copper loss is included, the 4-turn design is 91.3% efficient. Of course, this will depend on the wire gauge used. The electromagnetic properties of an antenna can be shifted from one frequency to another by multiplying all the dimensions by the (current frequency)/(target frequency). The 4-turn design can be scaled to 300 MHz by multiplying the radii and length by 229.33/300 = 0.7644. When this is done, exactly the same input impedance is found, but when loss is added, the efficiency is slightly less, 90.1%. This happens because, as mentioned in Chapter 2, skin depth and resistance scale as the square root of frequency.

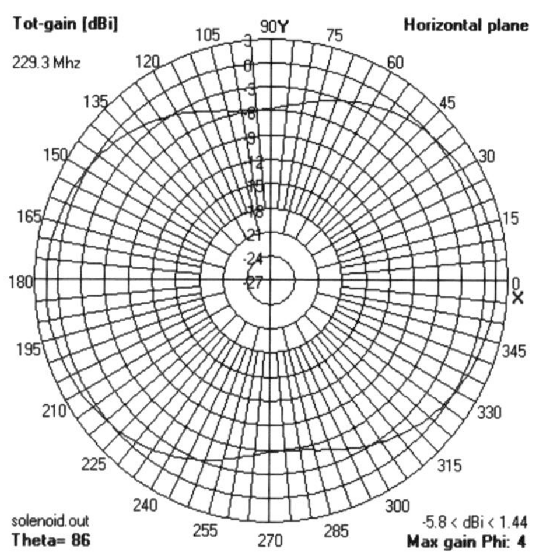

Figure 7.15: Horizontal radiation pattern for the 4-turn free-space solenoid of Table 7.7 and Figure 7.13.

7.3 The Contrawound Toroidal Helix Antenna (CTHA)

In the last section, we saw that a multiturn winding can be operated at a resonant frequency with reasonable Q and a pattern that has no null, but rather a dip. The CTHA carries this idea to an extreme [3]. It consists of two windings wound in opposite directions on a toroidal form. Figure 7.16 is an illustration showing the windings and expanded view of the source region. If you start at the top of the source and follow the wire to the right, it comes back to the source from the bottom left. The second winding then starts at the top left and comes back to the bottom right. If you look at the first half-turn of one winding, it lies over the last half-turn of the other winding, forming an elliptical loop. This is illustrated in Figure 7.17 where arrows are used to show the reference direction of current flow in each winding. These loops are the source of radiation for the antenna. Looking at the two figures, you can also imagine that the windings are like a twisted-pair transmission line that starts out to one side of the source, goes around the form, and comes back and connects to the source in a cross-wired fashion. It is a little more useful to think of the windings as two transmission lines leaving the source in opposite directions and tied together in a shorting manner at the back of the form. This mutual shorting means that the first resonance will be that of a shorted line, a high input impedance with a current maximum at the back of the form. The second resonance has current maxima at the back of the form and at the input, the third resonance has three current maxima spaced 120° around the form with one at the back and a high input impedance. All resonant current patterns must have a maximum at the mutual short at the back of the form.

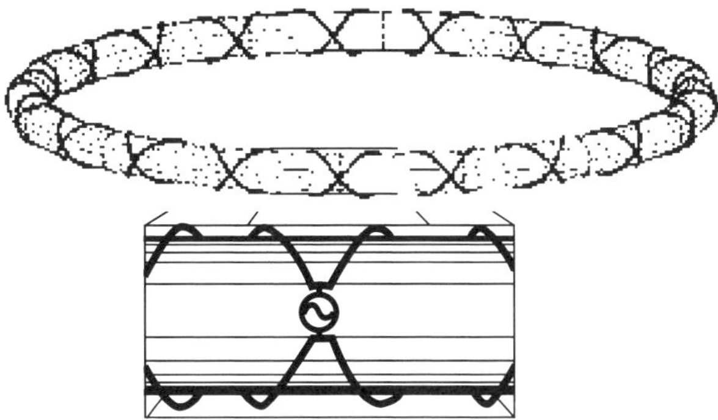

Figure 7.16: An illustration of the CTHA. The windings do not touch at the crossovers.

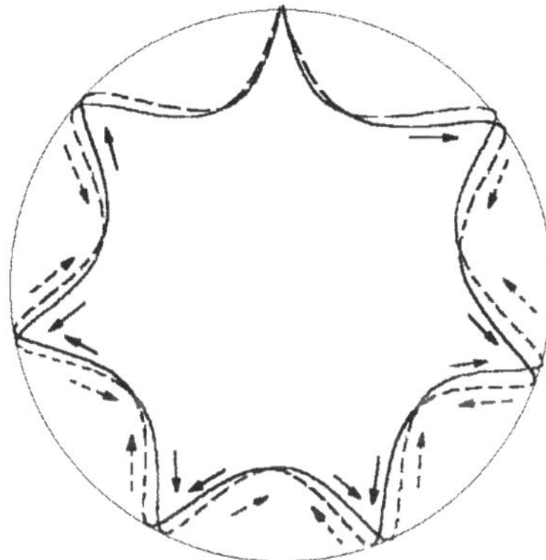

Figure 7.17: Top view of a CTHA winding. The arrows show reference current directions. Solid arrows and lines are on the top half, and broken arrows and lines are on the bottom half. Observe that each top half-turn of one winding lies over a bottom half-turn of the other winding, and the currents in each such pair are equal and opposite. This makes each such pair an equivalent loop with current circulating around the form.

It turns out that the radiation pattern at first or third resonance can have a quite small dip, and this is the main attractive feature of the CTHA. It has been controversial for the last 20 years or so, because its original developers didn't really understand its nature and so they built versions with too many turns. Faulty analysis and measurements led to exaggerated claims of its utility. For our present purpose, I have written a wire-list generator and associated files labeled ctha.*. Figure 7.18 illustrates the model of a 6-turn CTHA. In the code, the form's major radius is a and its minor radius is b. The wire radius is r. The space between the windings at the crossovers is also a parameter of the model. Getting a good APG figure was not easy for 2- and 3-turn versions. Details of the runs used to generate the following table are in ctha.txt on the CD-ROM. Since there are so many parameters to play with, a complete exploration of the performance as a function of the geometry would take more space than I want to devote. Instead, I offer the results for a particular case of a fixed form, wire size, and varying number of turns, as given in Table 7.8, which gives you a flavor of the type of performance to be expected.

Table 7.8: Results for increasing number of turns for designs with a = 0.5 m, b = 0.125 m, r = 0.0127 m, in free space. Copper loss is included. The overall diameter is $d = 2(a + b + 5r)$ = 1.377 and the height is $h = 2(b + 5r)$ = 0.377 m. d/h = 3.65. The gain dip is Dip = $G_{max} - G_{min}$ as found using a 5°-step grid over the quarter sphere.

First resonance.

n	*f, MHz*	*h/λ*	*R, kΩ*	*Q*	*E, %*	*Dip, dB*
2	35.805	0.045	163	912	85	8
3	29.073	0.0355	380	2448	52.7	5.6
4	25.518	0.0321	550	3154	40	4.3
5	21.579	0.0271	641	3738	27	4.43
6	17.95	0.0226	645	3638	15	6.4

Third resonance.

n	*f, MHz*	*h/λ*	*R, kΩ*	*Q*	*E, %*	*Dip, dB*
2	101.4	0.127	2.17	24	99.5	6.6
3	89.057	0.112	8.33	125	98.7	16
4	76.118	0.0956	15.19	229	97	20
5	65.162	0.0819	23.4	392	96	5.08
6	54.616	0.0606	30.96	533	93.5	4

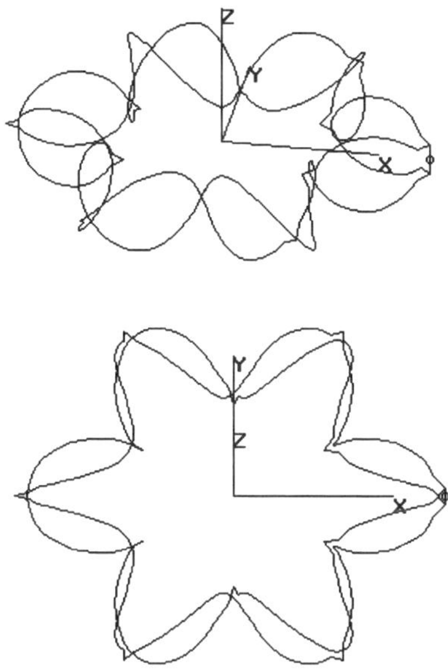

Figure 7.18: Two views of the NEC model of a 6-turn CTHA.

If you stack the efficiency and height/wave columns for the third resonance over those for the first resonance, they look rather continuous. Even with the large tubing used, the loss at first resonance for more than a few turns is generally unacceptable. Of course, the minor radius could be made larger to boost the radiation resistance. This has a tendency to make the dip larger as well, although one has to experiment with a particular design to see what the optimum might be. For third resonance design, since there are three current maxima it seems reasonable that you should get the best performance when you have turns, instead of crossovers, at the current maxima. This may be why the 6-turn case has the least dip and the 4-turn case the largest.

The data generation and analysis required some care because of the high Q. The impedance varies rapidly around the resonances, so one has to sample it over small frequency steps. Figure 7.19 shows a plot of the impedance for the 5-turn first resonance case. The data for this plot was generated at 250-Hz steps. In general, when searching for the resonant frequency I narrowed it down to a 1-kHz interval for which I had a positive reactance on one side and a negative reactance on the other. I then used datanec.m to find the resonant frequency, resistance, Q, and equivalent parallel L and C. The L and C are not instructive, so I didn't include them in the table, but for interest I did use them to find the impedance of the equivalent circuit over the same frequency range as the plot, and the curves were indistinguishable. Notice that the reactance slope is negative around the resonant point. There is a theorem in network theory that says the reactance slope of a lossless one-port network is always positive. This works for lossy series-resonant circuits as well, but not for lossy parallel-resonant circuits. These facts will be discussed further when we talk about measurements.

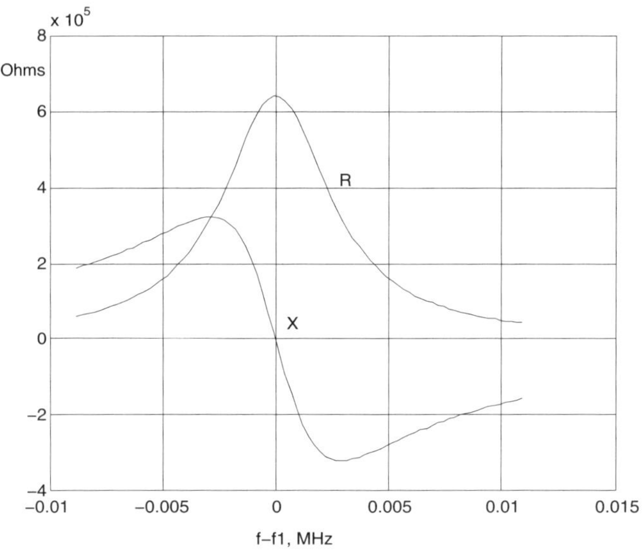

Figure 7.19: Input impedance for the 5-turn CTHA near first resonance (f1).

The CTHA is affected by the environment it's used in, just as any other antenna. Over a good ground plane, its resonant frequency is reduced. It is also reduced if the form is a dielectric core. The high resistance and *Q* at first resonance makes it very hard to tune or match. Since it is also a transmission line system, one can modify its current distribution by opening up the backside lines with either an open circuit, which inverts the current distribution and impedance characteristics, or a series load. Further, CTHAs could be cascaded to improve pattern stability and possibly have a good effect on impedance. These possibilities have not yet received serious study.

7.4 The Folded Spherical Helix Monopole

The sphere as an antenna shape has interested electromagneticians for decades. It has been proposed for dipoles and monopoles, and as a form for winding various coil antennas. Like the fat cylindrical monopole, feeding a hemisphere is problematic and I've never seen one in practice. However, as I mentioned at the end of Chapter 1, Steven R. Best has published a description of a class of coil antennas called Folded Spherical Helix antennas [4]. His paper includes both numerical modeling and some experimental results. These antennas have the potentially useful property of being resonant with a useful impedance when the size is border-line small. Best presents them in monopole version, and I have followed that lead by writing a set of files to model them over perfect ground, fsh.*. The FSH consists of a number of arms spiraling up from the ground plane to meet over their common center. One is driven, and the others act as shorted loads. Figure 7.20 shows the top view of a 5-arm version.

Figure 7.20: Top view of a 5-arm Folded Spherical Helix monopole wire model. The source is at the *X*. The turns appear more crowded at the outside because they are climbing up the steepest part of the hemisphere.

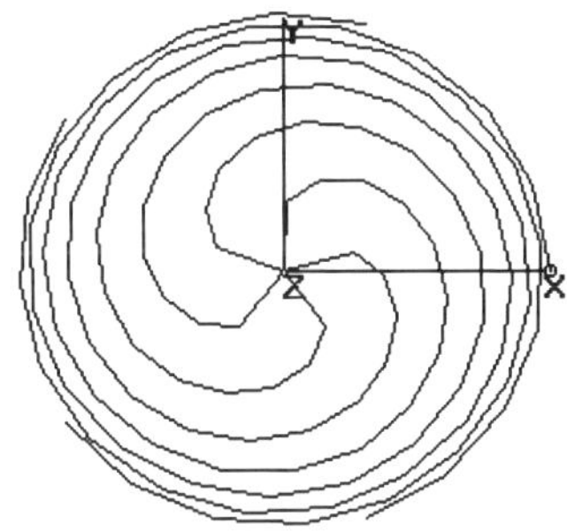

Best gives examples with even numbers of arms that show the resonant frequency, impedance and $1/Q$ go up with increasing arm number, and go down with increasing arm length for a fixed sphere radius. Table 7.9 shows the effects of changing the number of arms.

Table 7.9: Performance with different numbers of arms for FSH monopoles. Sphere radius is R = 60 mm, wire diameter is d_w = 1.6 mm, and copper loss is included. All cases are for 1 turn per arm. $H = R + d_w$ = 61.6 mm is height of the antenna.

Arms	*MHz*	*h/λ*	R_{in}, Ω	*Q*	*E, %*
2	281.2	0.0577	8.54	55	96.5
3	291.5	0.06	21.7	42	98
4	297.7	0.0611	43.35	36.4	98
5	301	0.0618	72.8	32.3	98.6
6	303.7	0.0624	119	31.8	99

As they stand, the 4-arm 1-turn case is close to 50Ω. A possible approach to getting a smaller design with a 50Ω input is to take a case with a higher R_{in}, shrink it to raise the resonant frequency, and then use more turns/arm to bring the resonant frequency and resistance down. I did this with the 5-arm case. I left the wire size the same, reduced the sphere radius and increased the number of turns/arm, aiming at 280 MHz and 50Ω. The result is R = 51.463 mm, turns/arm = 1.3,

$h/\lambda = 0.0495$, $R_{in} = 50.6\Omega$, $Q = 62$, and $E = 98\%$. The Q is somewhat higher than those for the volume-loaded dipole and the barrel loop, but this FSH doesn't need tuning. The half-power bandwidth is 9 MHz, and the SWR = 2 bandwidth is 3.2 MHz.

I used NEC2 to find the impedance vs. frequency and the resonant current distribution at resonance for the 5-arm 50Ω design. These are shown in Figures 7.21 and 7.22, respectively. Neither of these are exactly what one would expect, at least with casual thought. The resonance at 280 MHz is clearly present in Figure 7.21, but then there's a high-impedance resonance a little further on, rather than at a simple multiple. The current plot helps to explain this. We see that the pattern on each arm is more than a quarter cycle of a wave. As frequency is raised, this pattern doesn't have far to go to become a half-cycle, producing the high impedance. A curious feature of the current plot is that the imaginary part of the current is not close to zero everywhere as it is in simpler antennas we have studied.

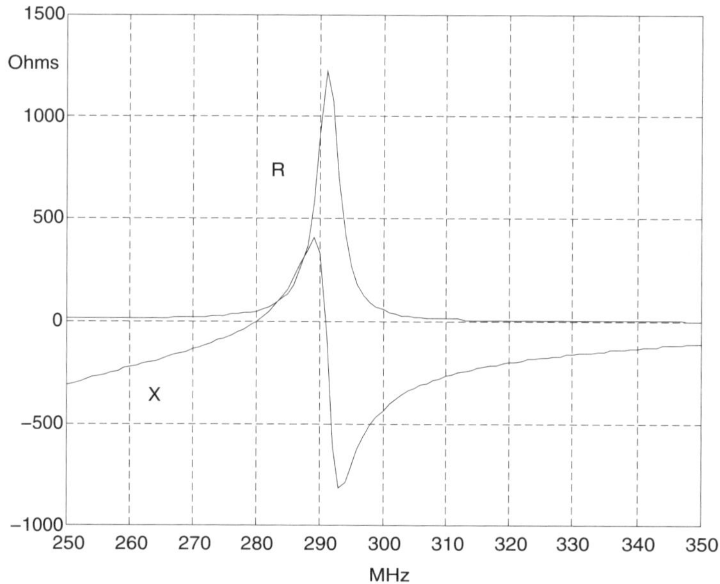

Figure 7.21: Input impedance for the 5-arm 1.3 turns/arm FSH monopole.

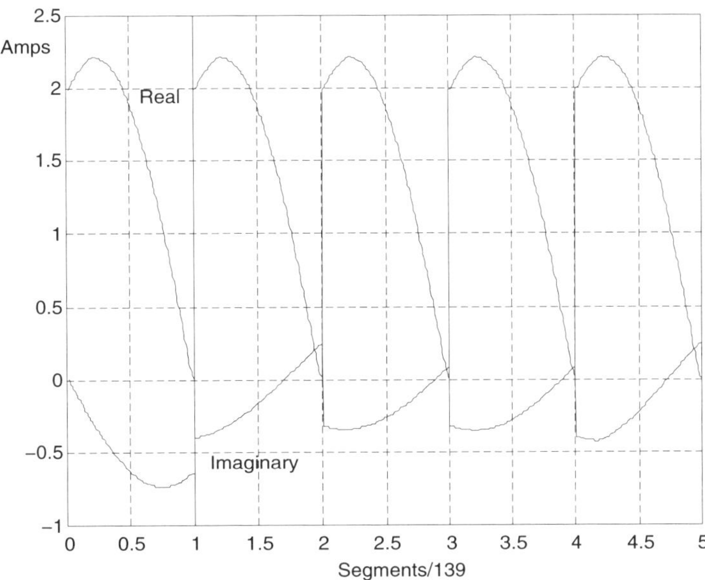

Figure 7.22: Current along the 5 arms of the resonant 280-MHz design. Each of the five sections is a graph of the current in one arm going from the ground plane to the antenna top. The leftmost graph is for the driven arm, and the following are for the arms in increments of 75° around from the X axis. The source is 100V.

7.5 Final Comments

The basic loop is a useful complement to dipole-type antennas because of its dual pattern. The barrel loop is its most efficient and lowest-Q form. The resonant solenoid offers an easier impedance-match at the cost of higher Q than the barrel loop. In recent years, many complex compact antennas have been proposed. The CTHA is an example that came to market too soon. The FSH is an interesting new antenna whose application potential is unrealized at present.

References

[1] *Reference Data for Engineers: Radio, Electronics, Computers, and Communications*, 7th ed., 2nd printing, Howard W. Sams, Indianopolis, 1985, p. 6-2.

[2] Glenn S. Smith, "Loop Antennas," note 2 on p. 5-22, Ch. 5 in *Antenna Engineering Handbook*, 3rd ed., R. C. Johnson ed., McGraw-Hill, 1993.

[3] D. B. Miron, "A Study of the CTHA Based on Analytical Models," IEEE Trans. on Antennas and Propagation, August, 2001.

[4] S. R. Best, "The Radiation Properties of Electrically Small Folded Spherical Helix Antennas," IEEE Trans. on Antennas and Propagation, vol. 52, no. 4, pp. 953–960, April, 2004.

Chapter 7 Problems

Section 7.1.1

7.1 Compare two single-turn loops at 14, 21, and 28 MHz. One is a circle of 2 m diameter and a square 2 m on a side. Both are made of aluminum tubing 25 mm in diameter. Use NEC single-wire models to compute impedance, Q, efficiency, and SWR = 2 bandwidth in free space. Prepare a table of the results and comment on their practical consequences.

7.2 Repeat problem 7.1, except that the circular loop is 1 m in diameter and the other loop is a rectangle 1 m by 2 m.

7.3 With Table 7.2 as a starting point, exercise the donut.* programs to find a/b for minimum Q.

7.4 Find a/b for minimum Q when $D = 0.05\ \lambda$. Find the efficiency and Q when copper loss is added.

7.5 For $a/b = 0.1$ and $D = 0.1\ \lambda$, find and plot the currents flowing around the shortest and longest paths that include the source.

7.6 For $a/b = 0.1$ and $D = 0.1\ \lambda$, find and plot the current that flows around the tube at its cross-section farthest from the source.

7.7 Rework Example 7.1 for average ground, $\sigma = 5$ mS, $\epsilon_r = 13$.

7.8 Rework Example 7.1 for good ground, $\sigma = 10$ mS, $\epsilon_r = 20$.

7.9 The loop in Example 7.1 can be resized to make its input resistance a good number at some useful frequency. For example, it is customary to operate a loop in a balanced mode with the source voltage equivalent to two equal voltages above and below ground. A transformer with three equal-turn windings can be used to convert a 200Ω balanced load to a 50Ω single-ended load, which is what your transceiver wants to see.

(a) Resize the loop, keeping $a/b = 0.1$, so that $Z_{in} = 200 + jX$, $X > 0$, at 28.9 MHz over average ground.

(b) Sketch a transformer and capacitor arrangement to match the loop to 50Ω.

(c) Calculate the voltages and currents in your network of part (b) for 100W input. You may assume the transformer is ideal.

(d) Rework Table 7.4 for the resized loop over average ground.

Section 7.1.2

7.10 Develop a 4nec2 program using SY commands to model two loops connected in parallel. Use two vertical single-wire circles that don't quite close, two horizontal wires to connect them in parallel, and a wire joining the centers of the horizontal wires for the source. It should look like the outline of the barrel loop in Figure 7.3. Make all dimensions and the segment density variable. For $f = 30$ MHz, $D = 1$ m, $r_w = 1.6$ mm, calculate and plot Q against loop separation over a range sufficient to show either a minimum or diminishing returns.

7.11 For the $N = 58$ entry in Table 7.5, find the current at a circular edge and the current in the middle circle of the barrel and plot them on the same graph. Other than at the feed gap edges, are the currents on the straight wires significant?

7.12 For the cell phone of Example 7.2, what is the SWR at 820 and 880 MHz?

7.13 For the cell phone of Example 7.2, design an impedance-matching network using two capacitors. The desired input impedance is 50Ω. Calculate the SWR at 820 and 880 MHz.

7.14 Rework problem 7.13 for the cell phone 1.7 m above average ground, $\sigma = 5$ mS/m, $\epsilon_r = 13$.

7.15 Is anything gained by making the loop in Example 7.2 wider—that is, occupy more of the box top? If so, what?

7.16 For Example 7.3, plot the radiation patterns for each polarization for the radio vertical in free space.

7.17 For Example 7.3, with the radio oriented as in Figure 7.12, choose two segments near the center of the strap that are perpendicular to each other, and find the currents induced by incident plane waves as illustrated in the figure. Compare the results to Table 7.6 and comment.

Section 7.2

7.18 Consider a 2-m length of insulated #14 wire. Assume the insulation is 0.5 mm thick so that the overall wire diameter is about 2.6 mm. At 15 MHz it is 0.1λ long. Find the number of turns it takes to make a coil of length equal to its mean diameter with the turns touching. This may be done by writing a program or using reasonable analytic assumptions. Find the coil's Q neglecting loss. Compare this to a single-turn loop made of the same piece of wire.

7.19 From Table 7.7, make a new table whose columns are n, $\lambda_2 (= c/f_2)$, l_w/λ_2, $v_{sol} (= f_2 l_c)$, and $v_w (= f_2\, l_w)$. Comment on the trends.

7.20 For the free-space entries in Table 7.7 calculate the solenoid volume in cubic wavelengths at the second resonance. Plot Q_2 against this volume. Find a reasonable curve-fitting equation.

7.21 Rescale the first 4-turn solenoid to resonate at 150 MHz over perfect ground. Use the smallest gauge copper wire for which the efficiency is at least 95%. What is the SWR = 2 bandwidth?

Section 7.3

7.22 For the 6-turn CTHA of Table 7.8, plot the normalized current in one winding at third resonance. Then delete the windings' middle segment (opposite the source, on the back of the form) and plot the current again. Having forced a current null in one winding, what happens to the current in the other winding? What's the new input impedance? Does the resonant frequency change?

7.23 Find the segment described in problem 7.22 and load it with a series capacitor. Find the value of capacitance that makes the input resistance 50Ω at the listed third resonance, 54.616 MHz. Then find the input series capacitance required to tune the antenna.

Section 7.4

7.24 Design a 6-arm FSH monopole to have $Z_{in} = 50\Omega$ at 150 MHz using #14 copper wire. Find the Q, efficiency, and SWR = 2 bandwidth of your design. What is its size in wavelengths?

7.25 One can approach the design problem for an FSH monopole from the other direction. That is, choose a base design with a lower impedance than desired and shorten the arms until the desired resonant impedance is achieved. Then rescale the design for the desired frequency. This will give an electrically larger antenna with lower Q than the starting antenna. Use the 3-arm version to get a 50Ω, 150-MHz design. Find the Q and SWR = 2 bandwidth. Is the antenna electrically small?

7.26 This is a project problem. The FSH monopole radiates like a simple vertical wire because it is mounted on a PEC plane which suppresses horizontally polarized fields. Write a program to model a dipole version of the FSH. Explore its impedance and radiation properties in free space. Develop a table similar to Table 7.9.

CHAPTER 8

Receiving Antennas

8.0 Introduction

In Chapters 2 and 3, I pointed out that an antenna is a transducer, so any antenna is capable of both transmitting and receiving operation. However, in many applications the antenna is not required to transmit. These applications include broadcast reception, direction finding, and spectrum monitoring. Without the need to transmit, the voltage or power-handling capability needed is much lower, which makes the hardware less expensive, and makes it possible to use materials that otherwise wouldn't stand the stress of transmitting. The receive-only application makes signal-to-noise ratio and consistent performance over very wide frequency bandwidth criteria that can be improved at the expense of impedance-matching (maximum power transfer).

The nature and amount of noise at a receiver's detector is a function of the type of amplification used, the noise generated in the receiver, and the electromagnetic noise picked up by the antenna. For AM and SSB receivers, the amplification is linear so that the effect of noise on the amplitude and timing of the voltage at the detector are important. In FM and keyed-signal (digital) receivers, only the noise effects on the timing (zero-crossings) of the voltage are important so long as there is enough signal+noise to drive the limiters (saturating amplifier stages) to full amplitude. In either case, the job of the antenna is to transfer enough of the external electromagnetic noise into the receiver so that it is at least as important as the receiver's own self-generated noise. This implies a necessity for some noise analysis of the antenna and input stage taken together. This is part of the presentation in following sections.

Wide operating bandwidth has been achieved in one of two ways. The first way is to use the antenna as a circuit element in the input stage. As mentioned in

Chapter 1, the multiturn coil antenna or loopstick is used as a fixed inductor with a shunt variable capacitor as part of the receiver's first-stage tuning and filtering design. The second way is to use a preamplifier with the antenna in a circuit designed to achieve a constant transfer from the wave's field strength to an output circuit variable, either voltage or current. This combination is called an *active* receiving antenna. There have been several commercial versions of electrically small active receiving antennas, some based on the single-turn loop and some based on a short monopole.

8.1 External Noise

Beginning in the 1940s, various agencies have been measuring and compiling radio noise data. The International Consultative Committee on Radio (CCIR) and its successor, the International Telecommunications Union-Radio (ITU-R) have published a sequence of reports putting this information in several formats and making corrections. The latest is called Recommendation ITU-R P.372, *Radio Noise*, issued in 2001 [1]. Below 30 MHz, the main radio noise source is lightning strikes. Since there is almost always a storm going on somewhere in the world and there are more of them towards the equator, the noise is present at all times and locations. However, there is a great deal of variability in any given location, between locations, and between seasons. Figure 8.1 (Figure 2 in [1]) shows lightning-caused noise (atmospheric noise) summarized as two probability curves for the frequency range 0.1 to 100 MHz. Other noise sources are summarized as average-value curves. The noise value is in terms of available power from a lossless short monopole over a perfect ground plane, regardless of how the value was actually measured.

$$F_a = 10\log_{10}(P_n/(kT_o bw)) \tag{8.1}$$

in which P_n is the available power in watts, bw is bandwidth in Hz, k is Boltzmann's constant $1.38\text{x}10^{-23}$ J/K, and T_o is 290 K. It is converted to a wave field strength value by:

$$E_n = F_a + 20\log_{10}(f) + 10\log_{10}(bw) - 95.5 \text{ dB}(\mu\text{V/m}) \tag{8.2}$$

with f in MHz.

Man-made noise in any given location can be due both to equipment outside the receiving room or vehicle, and to equipment in the room or vehicle. The dominant off-site man-made noise has been noise carried by power lines. Noise currents get on the power lines from the switching action of controllers for lights

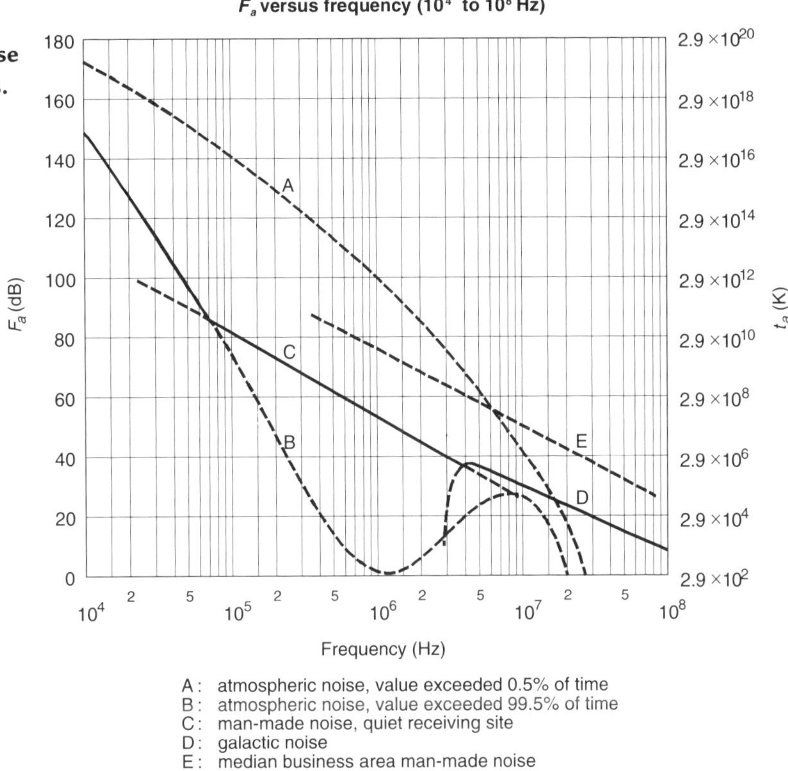

Figure 8.1 Radio noise from various sources.

and motors, and switching power converters. Defective items on or near the power line can also cause noise currents by arcing. At least one case was described in [2] in which two pieces of metal, neither of which was touching a power conductor, formed an arcing capacitor when the field strength got strong enough during the power cycle. The electromagnetic field of the arc current induced noise on the nearby lines, which then radiated noise. I have tracked down HF band noise from room light dimmer switches. The receiver was in a different building 20 m away. According to [2], as time goes by and more power lines are buried this source of noise will disappear in many locations. However, other measures must be taken to reduce noise from switching circuits in the same location as the receiver. Mitigation is discussed in [3].

8.2 The Ferrite Rod Antenna

Magnetic materials such as iron and its alloys are also good electrical conductors. If nothing is done to prevent it, applying an alternating magnetic field to a structure of magnetic material will induce large electrical currents and power loss.

At audio frequencies, transformers and motors have their magnetic parts built up from plates insulated from each other. The plating is oriented so that the plate lies in the desired magnetic field path and the thin direction is the one for the induced current. The conducting loops are then much smaller, enclose less magnetic field, and so the losses are much less. The loss increases with frequency, however, so something else has to be done to use magnetic materials at radio frequencies. The solution has been to grind the magnetic material into powder and separate the grains in a ceramic mixture. The result is called a ferrite. A great many formulations have been put on the market. In general, the higher the desired operating frequency, the lower the relative permeability one gets for an acceptable loss factor. Even though the magnetic material is reduced to grains, it still has some eddy current, and also it takes energy to flip those atomic magnets back and forth. These losses are represented by making the relative permeability a complex number.

$$\mu_r = \mu' - j\mu'' \approx \mu_r(1 - \mathrm{j}\tan\delta) \tag{8.3}$$

in which $\tan\delta$ is called the loss tangent. Both parts of μ_r are strong functions of frequency and also depend somewhat on the amplitude of the applied magnetic field.

8.2.1 Antenna Parameters

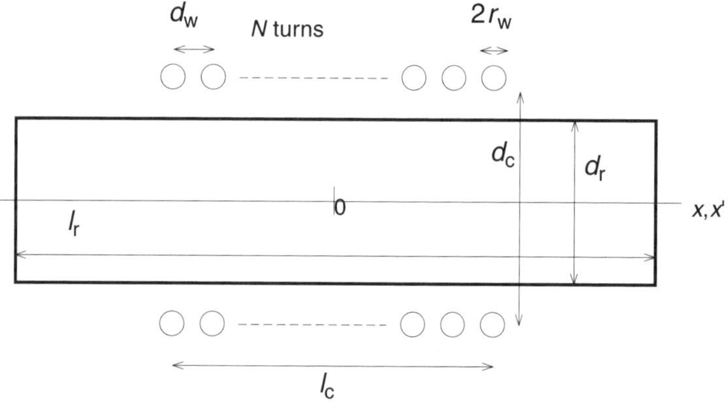

Figure 8.2: Longitudinal cross-section of a ferrite rod with *N* turns centered on it. Subscript "c" is for coil, "r" is for rod, and "w" is for wire. Note that d_w, d_c and l_c go between wire centers.

The ferrite rod antenna, known by various names including the loopstick antenna, was illustrated in Chapter 1. It generally consists of many turns of thin wire wound in a solenoid on a rod of ferrite material. The whole structure is very small electrically. There is usually an insulating material layer between the rod and the solenoid, and the wire may be bare or insulated, and the entire assembly may be dipped or coated for mechanical stability. There are a great many variables to choose, and my purpose in this section is to give you some analytical and numerical tools to help in this process. Figure 8.2 illustrates the basic dimensional variables and notation I use in the analysis.

We need to know the usual things about this antenna: the open-circuit voltage, the radiation resistance, loss resistance, and reactance. Because it is a series of small loops whose magnetic field is multiplied by the ferrite core, we may assume that the far-field pattern is the same as for the simple loop and that the directivity is 3/2. To find the inductance and the open-circuit voltage from a wave, we assume that the magnetic field in the rod at a transverse cross-section is x-directed, uniform over the area, and equal to the value on the axis. When a wave is present with its magnetic field pointing in the x direction, the maximum voltage will be induced in the coil. [4] describes an experiment in which a ferrite rod has a short coil ($l_c < d_r$) which is moved along it and the spot permeability is measured. It turns out that even with a uniform applied magnetic field, the voltage induced in the test coil drops as it is moved away from the rod center. For rods with small length/diameter (an example plot for $l/d = 21.4$ is given), the relationship is approximately:

$$\mu(x) = \mu_{\text{rod}}\left(1 - 3.6\left(\frac{x}{l}\right)^2\right) \tag{8.4}$$

in which μ_{rod} is the relative permeability measured at the center of the rod. μ_{rod} is itself a function of the length/width of the rod. The maximum value is that measured using a toroid core of the same material and it is the one usually designated as μ_r. Experimental curves have been given in many sources, most going back to [5]. [6] gives a generally used expression using the demagnetizing factor D, and a curve-fit expression good for $2 < l/d < 20$.

$$\mu_{\text{rod}} = \frac{\mu_r}{1 + D(\mu_r - 1)}, \quad D = 0.37\left(\frac{l_r}{d_r}\right)^{-1.44} \tag{8.5}$$

If one defines the point of diminishing return as the length for which $\mu_{\text{rod}} = \mu_r/\sqrt{2}$, the optimum length/diameter is

$$\left(\frac{l_r}{d_r}\right)_{\text{opt}} = 0.92(\mu_r - 1)^{0.69} \tag{8.6}$$

By averaging $\mu(x)$ from (8.4) over the length of a centered coil, we are on the way to getting an expression for the open-circuit voltage induced by a wave.

$$\mu_{\text{avg}} = \frac{\mu_{\text{rod}}}{l_c}\int_{-l_c/2}^{l_c/2}\left(1-3.6\left(\frac{x}{l_r}\right)^2\right)dx = \mu_{\text{rod}}\left(1-0.3\left(\frac{l_c}{l_r}\right)^2\right)$$
$$= \frac{\mu_r\left(1-0.3\left(\frac{l_c}{l_r}\right)^2\right)}{1+D(\mu_r-1)} \tag{8.7}$$

Now we are ready to find two of the antenna parameters. First, imagine a wave passing the antenna with an orientation such that the magnetic field vector, $\overline{\mathrm{H}}$, is along the x axis. From the expressions for the air-core solenoid given in section 7.2, we can now write:

$$V_w = h_e E_{\text{inc}} = N\mu_{\text{avg}}\beta A_r E_{\text{inc}} \tag{8.8}$$

and

$$R_{\text{rad}} = 20\mu_{\text{avg}}^2\left(NA_r\beta^2\right)^2 \tag{8.9}$$

in which A_r is the area of the rod's cross-section. The β in these expressions is still the free-space value, $2\pi f/c$.

Next, imagine that a current I is applied to the solenoid. This will produce an $\overline{\mathrm{H}}$ field that is nonuniform along the axis, and the rod's response to this field is even more nonuniform. There is no analytical help in determining the inductance and reflected ferrite loss, but [7] has a good deal of graphical data from experiments. Figure 4.12 [7, p. 190] shows that measurements of a kind of normalized inductance as a function of $l_n = l_c/l_r$ are grouped fairly closely. I did a quadratic fit to the central values in this graph and got the following expression:

$$L = \mu_{\text{rod}}\frac{N^2A_r}{l_r}F_L,\quad F_L = 10^{-3}\left(1.383l_n^2 - 3.518l_n + 3.08\right)\ \ \mu\text{H/mm} \tag{8.10}$$

The total back voltage of the coil due to the rod-and-coil interaction is

$$V_{\text{ind}} = \left(R_f + j\omega L\right)I \tag{8.11}$$

in which R_f is a resistance to account for the core loss. From electromagnetic theory, the power loss per unit volume at a point is

$$p = \frac{\omega}{2}B_oH_o\tan\delta = \frac{\omega\tan\delta}{2\mu_o\mu_r}B_o^2 \tag{8.12}$$

where B_o and H_o are the amplitudes of sinusoidal functions. I can't relate these directly to the coil current, but Figure 4.11(a) [7, p188] gives some curves for the normalized magnetic force field, $B(x)/B_c$, where B_c is the central (but unknown) force field. Following a sequence modeled in [7, pp. 189–190] the total power loss for the core is:

$$P = R_f I^2/2 = \int \text{pdvolume} = \frac{\omega A_r \tan\delta}{2\mu_o\mu_r}\int_{-l_r/2}^{l_r/2} B^2(x)dx$$

By assuming the loss is small, we may say approximately the $|I| = |V|/(\omega L)$ and use the voltage relation to the average force field, $V = \omega N A_r B_{\text{avg}}$. These substitutions lead to:

$$R_f = \frac{\omega L^2 \tan\delta}{\mu_o\mu_r A_r N^2 B_{\text{avg}}^2}\int_{-l_r/2}^{l_r/2} B^2(x)dx \tag{8.13}$$

This expression has the advantage that the B expressions can be replaced by normalized versions and the unknown B_c will drop out. The normalized data is given in the form $B_n(x_n)$ where $x_n = 2x/l_r$. This means that in the integrals, $dx = l_r dx_n/2$. Also, it's convenient to define a loss tangent for the rod,

$$\tan\delta_{\text{rod}} = \frac{R_f}{\omega L} = \frac{\mu_{\text{rod}} F_L l_n^2 \tan\delta}{\mu_o\mu_r}\cdot\frac{\int_0^1 B_n^2(x_n)dx_n}{\left[\int_0^{l_n} B_n(x_n)dx_n\right]^2} \tag{8.14}$$

A function-fit for Figure 4.11(a) of [7] that works pretty well is:

$$B_n(x_n) = 1 - \left[(1+d_1)x_n^2 - (d_1+d_2)x_n^{p_2} + d_2 x_n^4\right]\left(1 - l_n^{p_1}\right) - 0.85 x_n^{p_2} l_n^{p_1} \tag{8.15}$$

The ideas in this fit are that all the curves have zero slope at $x_n = 0$, and the plots change from a strongly curving one for $l_n = 1$ to a fairly straight one for $l_n = 0.1$. The constants are $p_1 = 0.65$, $p_2 = 2.6$, d_1=8.2, and $d_2 = 2$. The integrals in (8.14) have been evaluated and are embodied in the program ferrodloss.m.

We have the ferrite loss covered, but the copper loss is not just the usual skin effect. Because the wire is close-wound, the currents in adjacent wires crowd each other out-of-round, forcing more current into the side of each wire away from the group center. This is called proximity effect and it increases the copper loss because the current density in some parts of the cross-section is higher than the average implied by the total current. A method of calculating the increase was published in 1972[8] and a slightly simpler solution method is summarized in Appendix D. It is implemented in proxeffect.m. The copper loss is now written as:

$$R_c = \frac{l_w}{2\pi r_w}\sqrt{\frac{\pi f \mu_o}{\sigma}}\left(1+\frac{R_p}{R_o}\right) \approx \frac{N d_c}{2 r_w}\sqrt{\frac{\pi f \mu_o}{\sigma}}\left(1+\frac{R_p}{R_o}\right) \tag{8.16}$$

in which $l_w = N\sqrt{(\pi d_c)^2 + d_w^2}$ is the wire length and R_p/R_o is the ratio of the additional loss due to proximity effect to the loss without it.

Finally, we should make an estimate of the turn-to-turn capacitance. Borrowing and adapting the expression for two parallel wires (6.16), we have:

$$C_t = \frac{\pi^2 d_c \in_o \in_r}{\cosh^{-1}\left(\frac{d_w}{2r_w}\right)} \quad \textbf{(8.17)}$$

$\in_r$ is included in case one knows the values for the insulation and coating materials. In that case, one could use an average based on the volumes occupied between the wires. Since the *N* turns are in series, the total capacitance is:

$$C = \frac{C_t}{N-1} \quad \textbf{(8.18)}$$

Table 8.1 shows calculated results for a few cases illustrating performance vs. design parameters; *h* is the effective height in mm and *Qh* is the effective height of the resonated coil in mm. *Qh* is sometimes called a sensitivity and I use it here as a figure of merit for the loopsticks. Radiation resistance is not listed because it is in the order of μΩ for all cases. Table 8.1(a) shows that there is an optimum coil length that maximizes *Qh*, all other parameters being fixed. This happens because the average magnetic force field covered by the coil decreases with increasing length, a negative effect, and while the magnetic loss also decreases, the copper loss decreases more sharply as the proximity effect is decreased sharply with initial spreading of the turns, but is more stable at the longer coil lengths. Table 8.1(b) shows that one does get increases in *L* and *h* that are almost proportional to N^2 and *N*, but not quite, and *Q* goes down. This is again largely due to the reduction in average magnetic force field as more of the rod is covered. Choosing a ferrite with half the loss tangent cuts R_f in half, as expected, and therefore the *Q* in 8.1(c) is nearly double the value for the same geometry in 8.1(a). Doubling permeability and diameter both increase *Q*, but not to the extent one might intuitively expect, because both of these moves increases the core demagnetizing effect.

Table 8.1: Computed performance variation with design parameters. Initially, the rod is 203 mm long, 9.5 mm diameter. μ_r = 200, tanδ = 0.02. The coil diameter is 10 mm and the wire radius is 0.2 mm.

(a) Variation with winding length. The coil has 40 turns.

l_c, mm	Q	Qh, mm	h, mm	L, μH	R_f	R_w
18	107	667	6.25	164	7.61	2.04
20	116	722	6.25	162	7.57	1.23
25	121	753	6.24	157	7.48	0.7
30	121	753	6.22	153	7.39	0.535
35	120	746	6.21	148	7.29	0.461

(b) Variation with no. of turns. Coil length is chosen at max. Qh for a multiple of 10 mm.

N	l_c, mm	Q	Qh, mm	h, mm	L, μH	R_f	R_w
60	40	118.5	1101	9.3	324	16.2	0.984
80	50	116	1425	12.3	542	27.9	1.52
100	60	113	1731	15.3	796	41.9	2.13

(c) Variation with ferrite loss, tanδ = 0.01, N = 40, l_c = 20 mm.

Q	Qh, mm	h, mm	L, μH	R_f	R_w
203	1267	6.25	162	3.79	1.23

(d) Increase μ_r to 400.

Q	Qh, mm	h, mm	L, μH	R_f	R_w
293	2480	8.47	220	3.48	1.23

(e) Increasing rod and coil diameters to 12.7 and 13.2 mm.

Q	Qh, mm	h, mm	L, μH	R_f	R_w
361	4099	11.35	294	3.5	1.62

8.2.2 Circuit Applications

Figure 8.3 shows the receiving antenna schematic with a general load and two possible tuning capacitors. Since the antenna is usually used as part of a tuned input circuit, one of these is used as the resonating capacitor and the other is absent, depending on the type of load. The load might be the gate of an FET, essentially a capacitance, and the objective then would be to maximize the voltage. The load might be a finite resistance and reactance, such as the base or emitter of a BJT, and the objective might be to maximize the load (input) power. The load might be an effective short circuit, such as the summing junction of a feedback amplifier, and the objective would be to maximize input current. In the first case, parallel resonating the coil maximizes the voltage. In the third case, series resonating the coil maximizes the current while providing selectivity. In the second case, both capacitors may be needed to provide impedance matching for maximum power into the transistor. Of course, the two-capacitor arrangement isn't the only way to achieve impedance transformation.

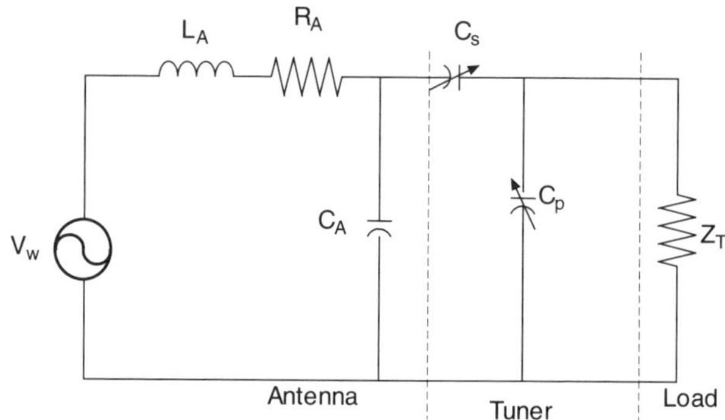

Figure 8.3: Equivalent circuit of the loop with tuning and a general load. $R_A = R_{rad} + R_f + R_c$.

Example 8.1 **A 75-m Band Receiver**

A narrow-band receiver designed to tune from 3 to 5 MHz is described in [9]. I thought it would be interesting to investigate the design of a loopstick antenna for it. At a certain point in the analysis, I found it would be good to transform down the tuned coil impedance to make it the same magnitude as the transistor's feedback capacitance reactance. So I added a second winding to the antenna, as shown in the schematic in Figure 8.4. [9] gives the receiver schematic, but not the bias-point values, so I had to estimate them. The data sheet for the 2N3819 gives the range of I_{DSS} as 2 to 20, so I chose the geometric mean, 6.3 mA. The data sheet gives a typical g_{mss} = 5.6 mS, so I used the FET transconductance expression, $g_m = 2(I_{DSS}/V_p^2)(V_{GS} - V_p)$ to find V_p = –2.25V. Next I used the static drain current equation and the source resistor voltage drop to find I_D = 1.36 mA, V_{GS} = –1.205V and g_m = 2.6 mS. I used a JFET noise model in [10, p. 306–307], and decided that the drain-current shot noise and flicker noise are probably the dominant components. However, I couldn't find enough information to calculate the flicker noise for the 2N3819, so I left it out. The small-signal equivalent circuit is shown in Figure 8.5 with the antenna tuned.

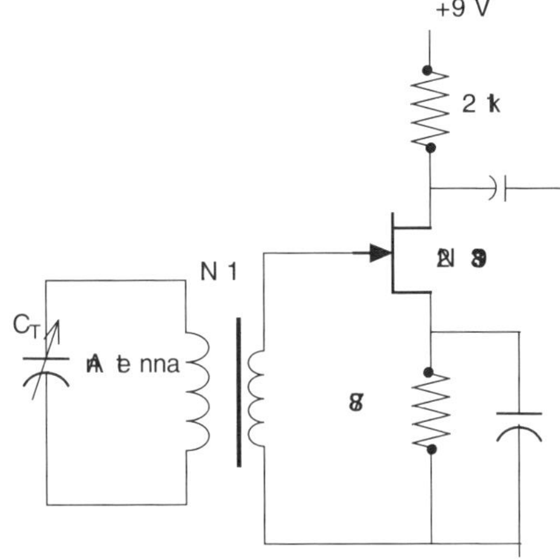

Figure 8.4: Receiver input stage and a two-winding tuned antenna.

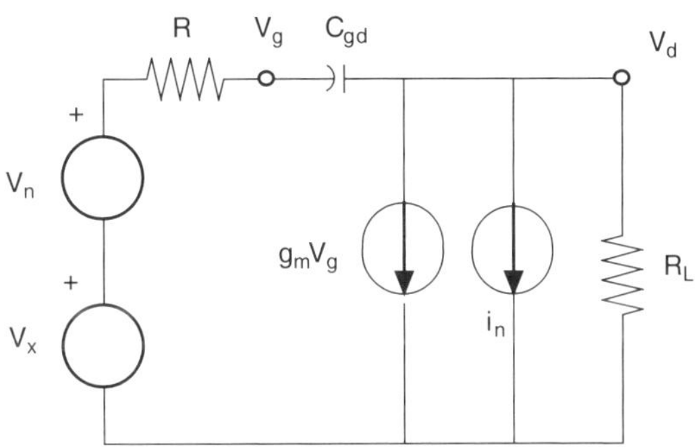

Figure 8.5: Small-signal equivalent circuit for Figure 8.4.

The elements in the small-signal equivalent relate to Figures 8.3 and 8.4 as follows: $V_x = (Q/N)V_w = (Q/N)hE_x$, $R = (Q/N)^2R_A$, $V_n^{\ 2} = 4kTRbw = (Q/N)^2V_{nr}^{\ 2}$ is the noise due to antenna loss, $i_n^{\ 2} = 8kTg_mbw/3 + 4kTbw/R_L$ is the current for the shot noise and the load resistor noise. Define $Y = j\omega C_{gd}/(1 + j\omega C_{gd}R)$, $G = 1/R$, and $G_L = 1/R_L$. Then circuit analysis leads to:

$$V_d\left(Y + G_L + YRg_m\right) = -i_n + \left(V_x + V_n\right)\frac{j\omega C_{gd} - g_m}{1 + j\omega C_{gd}R} \qquad \textbf{(8.19)}$$

The ratio of output power due to external sources to output power due to internal noise sources is:

$$\frac{S_x}{N_c} = \frac{V_x^2\left(\omega^2 C_{gd}^2 + g_m^2\right)}{i_n^2\left(1+\omega^2 C_{gd}^2 R^2\right) + V_n^2\left(\omega^2 C_{gd}^2 + g_m^2\right)}$$

$$= \frac{\left(\frac{Qh}{N}\right)^2 E_x^2\left(\omega^2 C_{gd}^2 + g_m^2\right)}{i_n^2\left(1+\omega^2 C_{gd}^2\left(\frac{Q}{N}\right)^4 R_A^2\right) + \left(\frac{Q}{N}\right)^2 V_{nr}^2\left(\omega^2 C_{gd}^2 + g_m^2\right)} \tag{8.20}$$

I designed the antenna coil to resonate at 4 MHz with a 30 pF tuning capacitor, an easily obtained value. That makes all the parameters in (8.20) known except the turns ratio, N. With $x = (Q/N)^2$, (8.20) has the form $f(x) = dx/(a + bx + cx^2)$, which has a maximum at $x^2 = a/c$. Then

$$\left(\frac{Q}{N}\right)^2 R_A \omega C_{gd} = 1 \tag{8.21}$$

and

$$\frac{S_x}{N_c} = \frac{h^2 E_x^2\left(\omega^2 C_{gd}^2 + g_m^2\right)}{2\omega C_{gd} R_A i_n^2 + V_{nr}^2\left(\omega^2 C_{gd}^2 + g_m^2\right)} \approx \frac{g_m^2 h^2 E_x^2}{2\omega C_{gd} R_A i_n^2 + g_m^2 V_{nr}^2} \tag{8.22}$$

The last approximation works for 4 MHz because ωC_{gd} = 17.6 µS. If we go to the kTbw forms for the noise sources, R_A becomes a common factor in the denominator as well. (8.22) becomes

$$\frac{S_x}{N_c} \approx \frac{g_m^2 E_x^2}{2\omega C_{gd}\left(\frac{2g_m}{3} + G_L\right) + g_m^2} \cdot \frac{h^2}{4kTbwR_A} \approx \frac{h^2 E_x^2}{4kTbwR_A} \tag{8.23}$$

(8.23) shows that h^2/R_A is a figure of merit for the antenna which is more important than Qh, at least in this application.

For a unit ratio of external noise to circuit noise, $E_x^2 = 4kTbwR_A/h^2$. From the material descriptions in [7], I chose B10 ferrite, $\mu_r = 130$, $\tan\delta = 0.003$ at 4 MHz, for a rod 100 mm by 12.7 mm. To resonate with 30 pF at 4 MHz, I need

$L = 52.77\ \mu$H. After some experimenting, I found that increasing the coil length increases h^2/R_A and decreases L. When L dropped below the target value, I increased the number of turns by 2 and got both higher h^2/R_A and too much L. Eventually this process failed to improve the result, so I stopped at 30 turns over 75 mm. The results are:

$$Q = 876$$

$$h = 9.987 \text{ mm}$$

$$L = 52.46\ \mu\text{H}$$

$$h^2/R_A = 66.3 \text{ mm}^2/\Omega$$

$$R_f = 0.912\Omega$$

$$R_w = 0.593\Omega$$

For external noise equal to circuit noise, $E_x^{\ 2} = 2.415E - 4\ (\mu\text{V/m})^2/\text{Hz}$. In a 1-kHz bandwidth, this corresponds to an rms. level of sqrt(0.2415) = 0.49 μV/m. From (8.2) $F_a = 47.3$ dB. From Figure 8.1 the range at 4 MHz is 20 to 66 dB. The noise floor for this antenna and amplifier is a bit high.

There are a number of things that could be done to improve the noise floor, and there are some questions about this circuit that remain to be answered. I'll leave all that for the problems.

8.3 Active Receiving Antennas

An active receiving antenna is not just an antenna followed by an amplifier. It is an antenna and circuit combination that takes account of the antenna's transfer and reactance characteristics in order to flatten out and broaden the overall transfer from wave field strength to output voltage. Two basic arrangements have been used and are illustrated in Figure 8.6. The feedback arrangement in Figure 8.6(a) has been used for both loops and whips, and the open-loop arrangement in (b) has been used with whips.

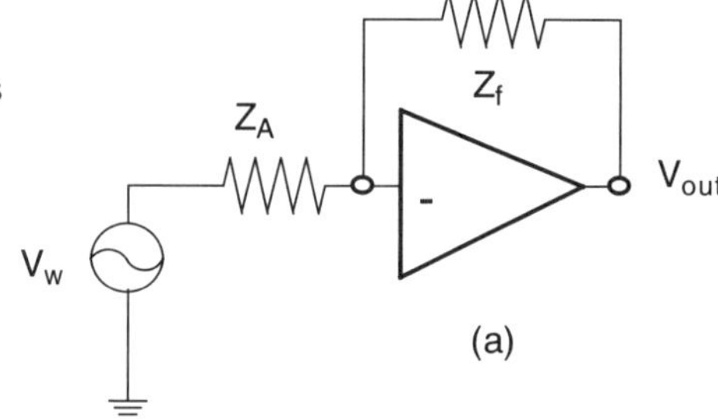

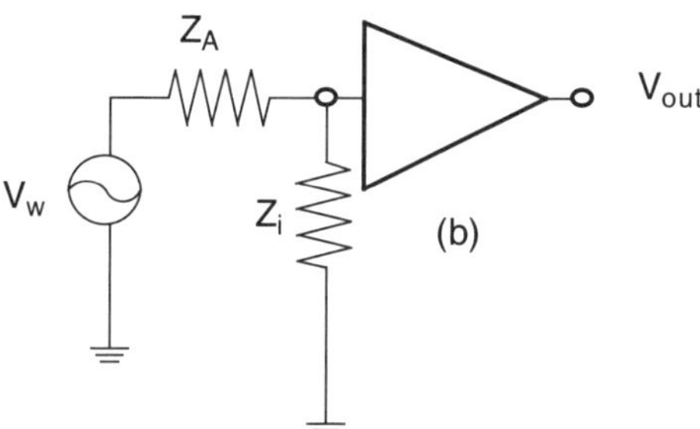

Figure 8.6: Two basic arrangements for active receiving antennas.

8.3.1 The Active Whip

For a short dipole or monopole, the transfer from field strength to open-circuit voltage is constant. The reactance is capacitive, $X_w = 1/(\omega C_A)$. In Figure 8.6(a), if the feedback element is also a capacitor, the gain will be independent of frequency. $A_V = -C_A/C_f$ and the overall transfer is $V_{out}/E_{inc} = h_e\ C_A/C_f$ In order to get an increased effective height, the feedback capacitor has to be smaller than the antenna capacitance. For a short, thin whip, this may be impractical, but it should be a reasonable approach for an end-loaded whip. In any case, the desired total gain can be made up with added amplifier stages.

For the arrangement in Figure 8.6(b) we choose $Z_i = -j/(\omega C_i)$ and we have a frequency-independent voltage divider. If A_V is the gain of the amplifier, the overall transfer is:

$$\frac{V_{out}}{E_{inc}} = \frac{A_V h_e}{1 + C_i/C_A} \tag{8.24}$$

Again, the smaller the antenna capacitance, the more amplifier gain has to be used to make up for the voltage division.

8.3.2 The Active Loop

As pointed out frequently in this book, the open-circuit voltage from a small loop is proportional to frequency, and so is the reactance. So, short-circuiting a loop antenna will produce a transfer from field strength to current that is independent of frequency. For a one-turn loop of area A and inductance L,

$$\frac{I_w}{E_{\text{inc}}} = \frac{A}{cL} \tag{8.25}$$

The feedback element in Figure 8.6(a) only needs to be a resistor to provide constant transfer to the output voltage.

$$\frac{V_{\text{out}}}{E_{\text{inc}}} = \frac{AR_f}{cL} \tag{8.26}$$

For an air-core case, if multiple turns are considered they should be in parallel rather than series to maximize the short-circuit current. With a ferrite core, this would probably still be true, but I've never seen a ferrite rod wound with a single wide-strap turn. Also, increasing the permeability increases both the voltage pickup and the inductance, so there may be little net gain.

8.3.3 General Considerations

The advantages of the active antenna are the constant transfer over a decades-wide frequency range, and an arbitrarily large signal from an arbitrarily small antenna. The first caveat you should think of, though, is noise. Just like the ferrite loopstick, the antenna element has to transfer enough external noise to mask the amplifier's noise and its own loss noise. The next issue is linearity. All amplifiers are built from nonlinear active devices, and the larger the signals they have to handle, the more nonlinear they become. Feedback internal to the amplifier can help this situation. The operating point of the first stage is a compromise between trying to keep the noise down and having enough bias current and voltage drop to handle fairly large signals without cross-modulating weak ones. The intended applications generally involve passing a wide frequency spectrum on to one or more narrow-band receivers that pick out the signal of interest to the operator at the moment. Harmonic generation and spurious mixing can occur with signals the operator isn't tuned to. As far as I can tell, only one of the commercial units currently available have filters. One unit has an amplifier bypass switch. The Hermes active loop sold in the last quarter of the 20th century had filters to block the AM and TV broadcast bands and passed 2–32 MHz.

Example 8.2 The Hermes Active Loop

The Hermes Active Loop was 1 yard in diameter and made of two parallel loops of 1" aluminum tubing. The inductance was about 1.4 μH and the loop-amplifier combination was designed to have an effective height of 1 m into a 50Ω load. Figure 8.7 shows a loop post-mounted with two air-speed coaxial lines connecting it into an array. Figure 8.8 shows a schematic of the first stage of a general amplifier, similar in intent and topology to the Hermes amplifier of the 1970s. All the voltage gain and the feedback are in the first stage. The second stage was an emitter follower, and the third stage was an open-collector (choke-loaded) common-emitter stage with an unbypassed emitter resistor to give unity voltage gain and very linear current gain into the RF load. The transformer T1 is 1:1, C1–C3 are bypass and blocking capacitors, and R1–R3 are determined by the bias requirements.

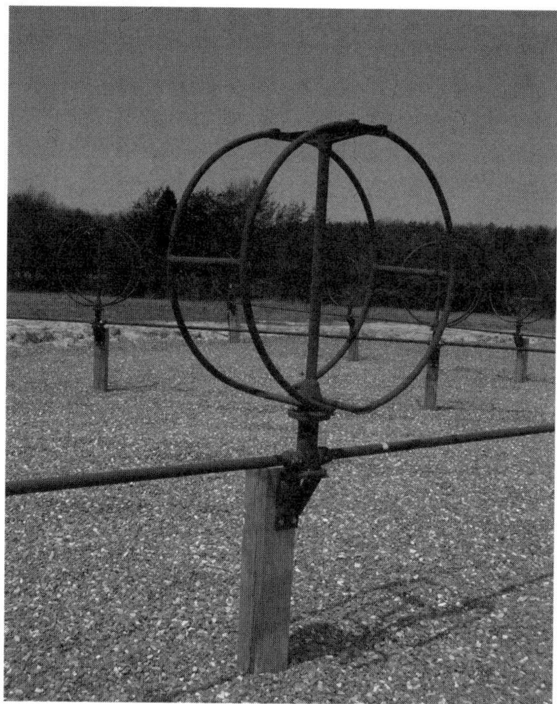

Figure 8.7: The Hermes Active Loop, post-mounted, with air-speed transmission lines to make it part of an array. Photo courtesy of US Antenna Products LLC, Frederick, MD, U.S., which is the current manufacturer.

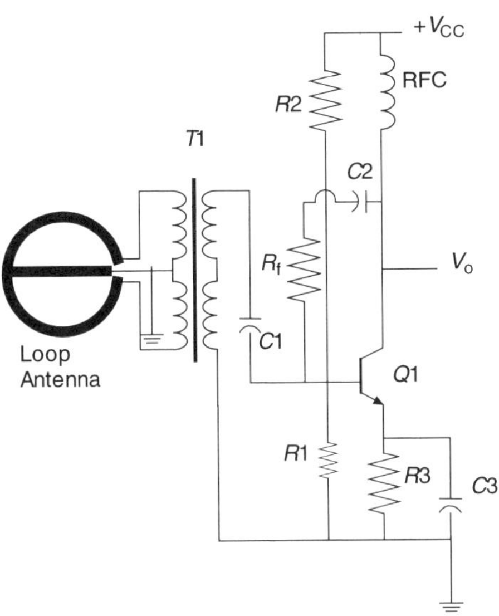

Figure 8.8: Schematic of a general version of the first stage of the amplifier for the Hermes Active Loop, circa 1977.

Figure 8.9 shows a small-signal equivalent circuit including two noise sources that depend on the base and collector bias currents.

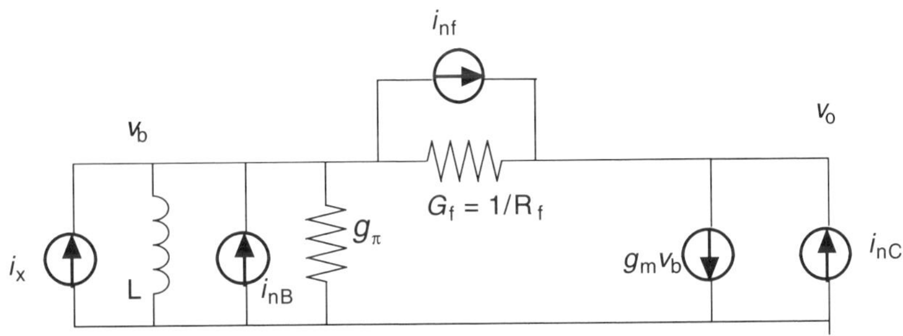

Figure 8.9: Small-signal equivalent circuit for Figure 8.8.

In Figure 8.9, I have neglected the transistor's internal feedback elements and the base bias resistors of Figure 8.8. i_x represents the current from external waves. The noise sources are given by $i_{nB}^2 = 2qI_Bbw$, $i_{nC}^2 = 2qI_Cbw$ [10, p. 305], and $i_{nf}^2 = 4kT_obw/R_f$. I_B and I_C are the base and collector bias currents. $q = 1.6\text{E} - 19\text{C}$ is the electron charge magnitude. Let $Y_b = g_\pi + 1/(j\omega L)$. Then:

$$v_o = \frac{1+R_f g_m}{g_m + Y_b}\left(i_x + i_{nB}\right) + \frac{1+R_f Y_b}{g_m + Y_b} i_{nC} - R_f i_{nf} \tag{8.27}$$

If g_m were a large number, v_o/i_x would go to R_f, the desired condition. Since R_f and L are already determined, the major design question is the choice of bias currents. Before doing any more algebra, let's take a look at some numbers. Let's choose I_C =1 mA and β = 100, which makes I_B = 1/100 = 10 μA. $g_m = (qI_C)/(kT_o)$ = 40 mS with T_o = 290 K. $g_\pi = g_m/\beta$ = 0.4 mS. At 2 MHz, 1/(ωL) = 56.84 mS and at 32 MHz 1/(ωL) = 3.55 mS. Even at 32 MHz $1/(\omega L) >> g_\pi$. $R_f g_m$ = 27.2 >>1. g_m and $1/(\omega L)$ are the same order over most of the frequency range. The ratio of output power from external sources to output power from circuit sources is approximately:

$$\frac{S_x}{N_c} = \frac{\left(R_f g\right)^2 I_C^2 i_x^2 / (2bw)}{\left(R_f g\right)^2 q I_C^3 / \beta + \left(\frac{R_f}{\omega L}\right)^2 q I_C + 2kT_o G_f \left(g^2 I_C^2 + \left(\frac{1}{\omega L}\right)^2\right)} \tag{8.28}$$

with $g = q/(kT_o)$. To find the value of I_C that gives maximum S/N requires solving a cubic equation, which is possible but not pretty or informative. So I wrote loopamp.m without the approximations used in (8.28) to calculate S/N, h_e, Fa, and E_x, the external noise field strength for S/N = 1 in bw Hz. Either I_C or frequency can be a vector. I found, of course, that the optimum collector current depends on the frequency. Specifically, at f = 2 MHz the minimum Fa = 45 dB at I_C = 14.1 mA, at 8 MHz minimum Fa=27.2 dB at I_C = 3.5 mA, and at 32 MHz minimum Fa = 9.4 dB at I_C = 0.9 mA. Since 8 MHz is the geometric center of the band, I chose I_C = 3.5 mA and generated the frequency responses for the performance parameters. Figure 8.10 shows Fa across the band, Figure 8.11 shows the effective height, and Figure 8.12 shows the noise floor for a 5-kHz bandwidth in μV/m. The Fa curve is quite close to the curve in Figure 8.1 for the minimum expected noise.

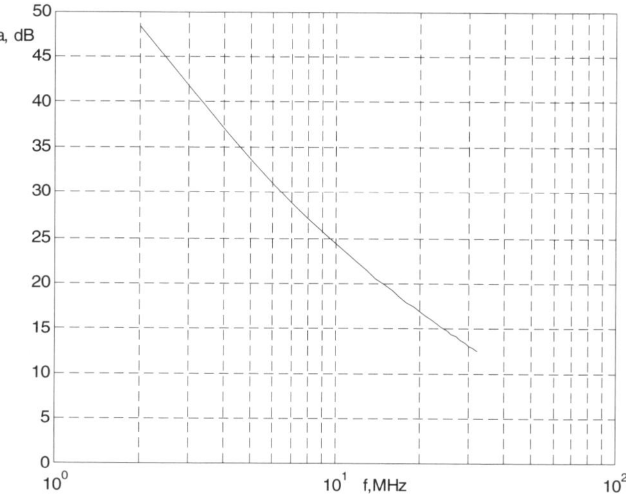

Figure 8.10: Equivalent CCIR noise factor for the loop and amplifier with I_c = 3.5 mA.

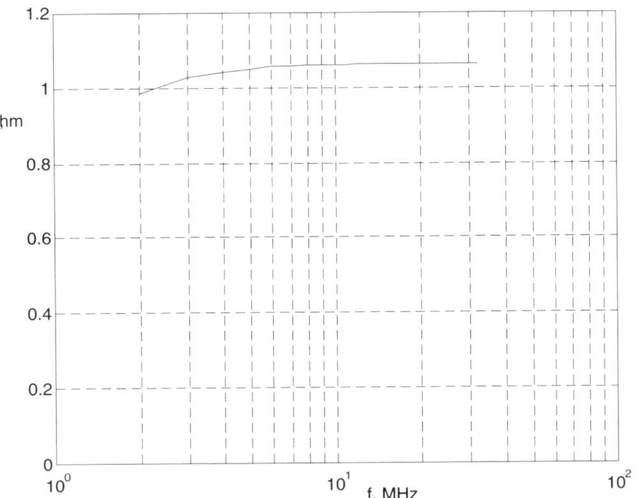

Figure 8.11: Effective height at I_C = 3.5 mA.

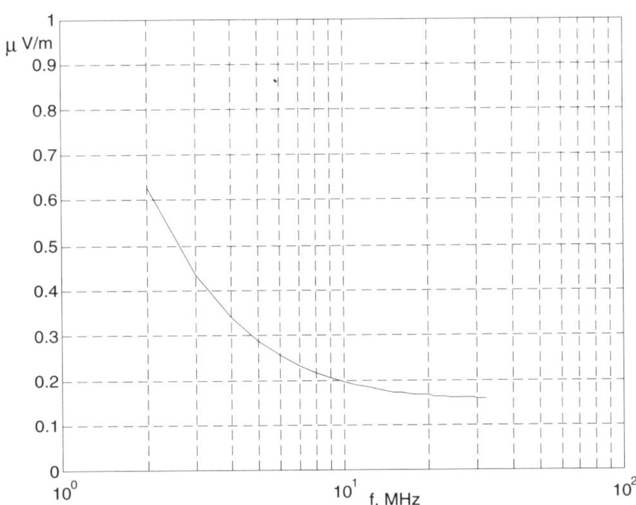

Figure 8.12: Loop and amplifier noise floor in terms of field strength in a 5-kHz bandwidth at I_C = 3.5 mA.

References

[1] *Radio Noise*, Rec. ITU-R P.372, ITU, Geneva, 2001.

[2] W. R. Vincent, R. W. Adler, and G. F. Munsch, "An Examination of Man-Made Radio Noise At 37 HF Receiving Sites," Naval Postgraduate School, Report NPS-EC-05-003, November 2004.

[3] W. R. Vincent and G. F. Munsch, "Power-Line Noise Mitigation Handbook," 5th ed., Naval Postgraduate School, Report no. NPS-EC-02-002, January 2002.

[4] Philips Tech. Rev., vol. 16, no. 7, pp. 191–192, Jan. 1955.

[5] R. M. Bozorth and D. M. Chapin, "Demagnetizing Factors of Rods," J. App. Physics, vol. 13, pp. 320–326, May 1942.

[6] R. C. Pettengill, et. al., "Receiving Antenna Design for Miniature Receivers," IEEE Trans. on Antennas and Propagation, vol. 25, no. 4, pp. 528–530, July 1977.

[7] E. C. Snelling, *Soft Ferrites Properties and Applications*, CRC Press, originally Ilifre Books, London, 1969.

[8] G. S. Smith, "Proximity Effect in Systems of Parallel Conductors," J. App. Physics, vol. 43, no. 5, pp. 2196–2203, May 1972.

[9] Bill Young, "A Cascade Regenerative Receiver," QEX, no. 222, pp. 7–11, Jan/Feb 2004.

[10] G. Massobrio and P. Antognetti, *Semiconductor Device Modeling with SPICE*, 2nd ed., McGraw-Hill, 1993.

Chapter 8 Problems

Section 8.1

8.1 Atmospheric noise has been described as $1/f$ noise. This is also true for the flicker noise in transistors. Convert a few points from Figure 8.1, curves A and B below 10 MHz, to field strength in μV/m for a 10 kHz bandwidth. Then find a constant for each set of values so that $e_n = k/f$ is a good approximation. Comment on the nature of the errors.

Section 8.2.1

8.2 Use eqn. (8.6) to make a graph of optimum length/diameter for $100 < \mu_r < 500$.

8.3 Verify (8.7) for the average permeability.

8.4 Derive (8.8) for the effective height of a loopstick.

8.5 Derive (8.9) for the radiation resistance.

8.6 Derive (8.14) for the rod loss tangent.

8.7 Derive an approximation for the copper loss in (8.16) that still contains d_w.

Section 8.2.2

8.8 Suppose the first antenna in Table 8.1 is to be used for an AM receiver. The receiver input is the 2500Ω resistance of a BJT base. Find the tuning ranges for the capacitors in Figure 8.3 to tune from 550 to 1750 kHz.

8.9 Use the arithmetic mean $I_{Dss} = 11$ mA in Example 8.1 and recalculate the pinchoff voltage and operating point values.

8.10 Verify the circuit analysis that led to (8.19).

8.11 Verify the power ratio expression in (8.20).

8.12 In Example 8.1 what is the value of N?

8.13 In Example 8.1 what is the value of V_d/E_x?

8.14 Which rod dimension would improve the performance of the antenna in Example 8.1 the most if it were changed?

8.15 In Example 8.1 set the core diameter to 25.4 mm and the coil diameter to 30.4 mm. Redesign the coil and find the new noise floor.

8.16 In Example 8.1 set the rod length to 203 mm and redesign the coil. Find the new noise floor.

Section 8.3.1

8.17 Derive the gain expression for a whip and feedback capacitor given in the first paragraph.

8.18 Suppose an active whip is designed in the feedback arrangement of the first paragraph. Let the no-load transfer function be $A(f) = A_o/(1 + jf/f_o)$ and let the output have a series resistance R_o. Analyze the system for stability.

8.19 Consider an arrangement as shown in Figure P8.19. With each of the Z_k single circuit elements, as opposed to being a network, it is possible to obtain a flat transfer without using a capacitor smaller than C_A. Find such a circuit and analyze it.

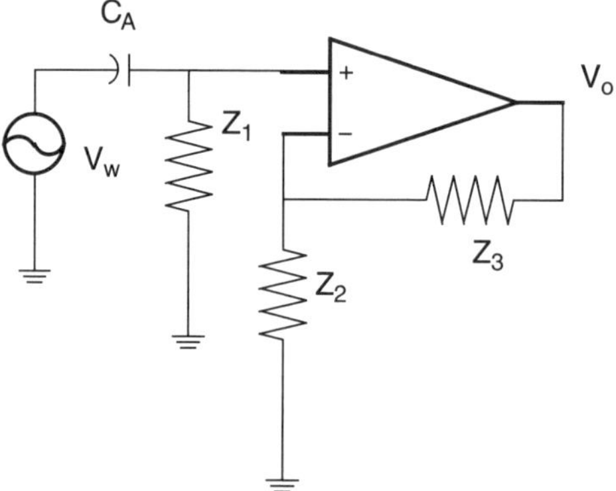

Figure P8.19.

8.20 Suppose a short dipole is connected to the 2N3819 FET amplifier of Example 8.1. Let the dipole be 2 m long and 3 cm in diameter. Do a gain and noise analysis for operation from 2 to 20 MHz. Assume that $C_{gs} = 2.3$ pF and $C_{gd} = 0.7$ pF. Suggest improvements.

Section 8.3.3

8.21 In the denominator of equation (8.28), it is reasonable to drop one term that is much smaller than the others. This will make it possible to find the I_C expression that maximizes *S*/*N*. Do so.

8.22 (a) Design the bias elements for Figure 8.8. Let the supply be 12 V. The maximum RF voltage will be set by $V_{CC}-V_E-V_{sat}$. This voltage divided by R_f should not exceed I_c. The RF choke acts like a constant-current source, so its current sets the limit on the RF current swing.

(b) Test your design in SPICE. Choose a high-frequency transistor, such as the 2N5109, for which your version has a model. Test harmonic generation by asking for a Fourier analysis for successively larger input signal amplitudes. Do some two-frequency intermodulation tests.

CHAPTER 9

Measurements

9.1 What Are You Measuring?

The meaning of this question is that, very often, people trying to measure some aspect of an antenna's performance are not seeing what they imagine they are measuring. Figure 9.1 illustrates a common problem with impedance and other measurements. The instrument making the measurement is necessarily at some distance from the antenna and is connected through a transmission line, usually a coaxial cable. Whether the load end is grounded or not, the mismatch causes a reflected wave, some of which leaves the feed region and causes a current wave on the outside of the coax. The fact that the reflection is not all in the cable causes a different impedance from that of the antenna itself, and the current on the outside of the cable radiates, making it appear that the antenna is sending out more power than it is by itself.

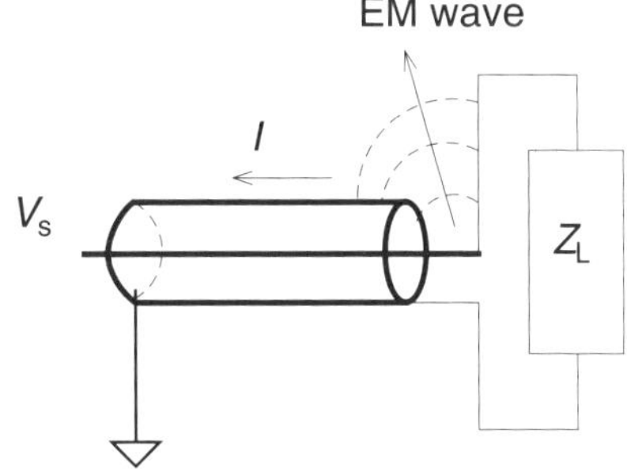

Figure 9.1: A mismatched physically open load produces a current on the outside of the coaxial cable.

Figure 9.2 shows a generic test range. For pattern and gain measurements, the antenna being tested (device under test, DUT) is in receive mode and a source antenna illuminates the DUT with a nominally plane wave. The DUT is mounted on a rotator capable of 3-D positioning and the transmitted field is stationary. Sounds simple and direct, right? But, as you see, there are possibilities for getting more waves than you want. Indeed, I was looking at a 20-year-old book I've had since it was new, and it describes a situation where both antennas are high-gain parabolic types which pass the energy back and forth between them without the kind of attenuation you'd expect from the far-field situation. Another problem with this setup is the size of the positioner relative to the antenna. It should either be small compared to the antenna, which is hard to arrange with electrically small antennas at higher frequencies, or the positioner can be effectively hidden by having a large ground plane on it. Arranging cables to minimize pickup and reradiation is another consideration.

In the following sections I will address some of the concerns raised above. In this chapter I will present only a small subset of the principles, ideas, and techniques involved in antenna measurements. I cover some of the items most relevant to small antennas and that reflect my personal experience.

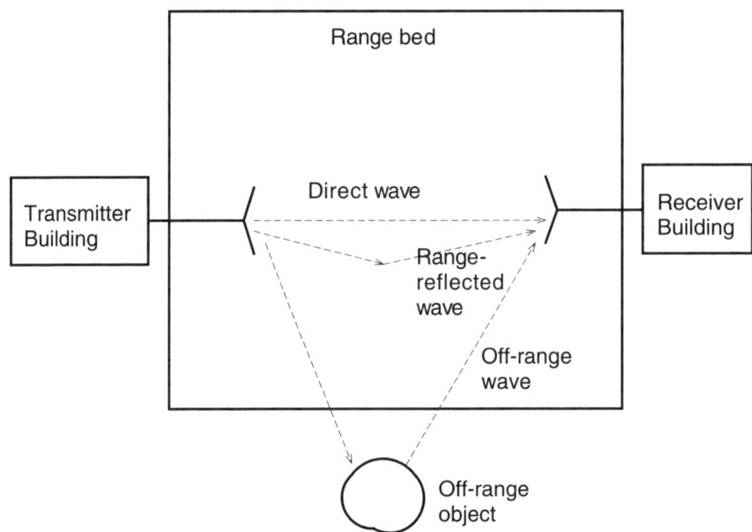

Figure 9.2: Generic antenna test range, showing some of the possible waves reaching the receiving antenna.

9.2 Measurements Through a Transmission Line

Nowadays one can buy SWR and impedance meters that are small and battery-powered [1]. They cover the HF band and portions of the VHF and UHF bands of interest to Amateur Radio Operators (hams). You can connect one of these directly to the antenna's feedpoint and make measurements without considering the feedline. However, especially at HF, you are standing in the near field of the antenna and making a difference to the measurement. You can back away from the antenna and watch the reading change, and keep moving until it stops changing. You may need a binocular to read the display, though. Also, the geometry of the meter leads and connection points may be different from that of the feedline connection, which will alter the local reactance components. For these reasons, it is best to make measurements through the feedline, especially the feedline that will actually be used with the antenna.

The issue of mismatch-caused radiating currents on the feedline can be dealt with in either of two ways. If you are not interested in an accurate measure of the antenna, but only in getting a good impedance match at its input, then working toward a good match at the feedline-antenna junction will minimize the effect. If you want a real measurement of the antenna's impedance or radiated power, you have to block that external current wave. One way to do this is with a *choke balun.* This is just a coil of the coaxial cable next to the connection point with the antenna. The series inductance of the coiled outer conductor can be made high enough to keep the wave going out the antenna or back inside the line. [2, p. 16–9] gives dimensions for some HF-band designs. Another approach is to slip ferrite toroids over the line near the antenna end, possibly several per quarter wave. The ferrite acts as a high impedance to the magnetic field of the coax surface current. Many people use a "hand test." If you can change the reading by grabbing the cable, there's an unwanted current on the sheath.

9.2.1 If I Only Have an SWR Meter...

...a variable-frequency signal source and a supply of well-defined components, what can I learn? I assume that from modeling the antenna and its test environment, you know that it is resonant in or near the frequency range of your signal source, which may be a generator, a transceiver with limited bands of operation, or contained in the SWR meter. To begin, use the following facts. For a series $R + jX$ load, the minimum SWR is at $X = 0$. When the load is seen through a cable, the SWR is the same, or lower if cable loss is significant. This means that when the SWR is minimized by varying the frequency, you can use its value to find the resonant resistance. The difficulty with this may be that the transmitter can't be tuned outside the allowed bands, so sometimes the best that can be done is to find the direction of decreasing SWR and make an adjustment accordingly. Consider the circuit in Figure 9.3. The antenna is shown as a resistance R with a series reactance jX. At a given frequency, the reactance may not be the value needed to form the series part of an L-section match. In that case, I have found that the minimum SWR that can be obtained by adding a single shunt susceptance, jB, is given by the following equations.

Define $Z_q = R^2 + X^2$.

Then for minimum SWR,
$$B = \frac{X}{Z_q}, \tag{9.1}$$

and
$$S = \frac{X}{RBZ_o}, \tag{9.2}$$

in which S = SWR and Z_o is usually 50Ω, the coax and system impedance. With two equations, two unknowns can be found. Knowing B and S, we find:

$$R = \frac{Z_o S}{1+\left(BZ_o S\right)^2} \tag{9.3}$$

$$X = RBZ_o S = \frac{B\left(Z_o S\right)^2}{1+\left(BZ_o S\right)^2} \tag{9.4}$$

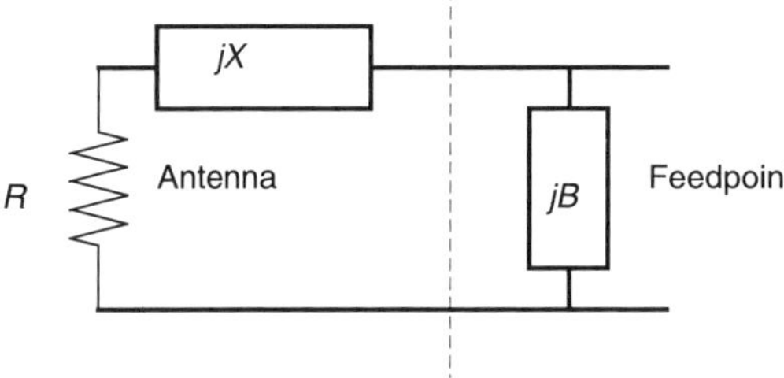

Figure 9.3: Transmit equivalent circuit for the antenna and one shunt tuning element.

Example 9.1 ***Volume-Loaded Monopole at 80 m***

In 2003, I designed and built a 4 m by 4 m volume-loaded dipole as an experiment [3]. The radiating element for 80 m was ¼" soft copper tubing about 13.4 m long and somewhat coiled. I adjusted the length to give a series resonance (minimum SWR) at a little above 3.5 MHz. For my tuning problem, I estimated a value of C to use, put it in, measured the minimum SWR and frequency, and found the antenna R and X from (9.3) and (9.4). For the coil, three 470-pF caps in parallel yielded SWR = 1.6, $Z_A = 9 + j25\Omega$ at 3.9 MHz. To tune at this frequency, the coil should be shortened to give $9 + j20\Omega$ and C should be 1660 pF. This was for the bare coil and connector, between the old relay boxes and the new ones. With the new boxes, I shortened the coil length and C = 1712 pF ±5% so that SWR≈1 from 3.94 – 3.96 MHz, ≈1.7 at 3.92 and 3.995 MHz, and = 2 at 3.9 MHz.

9.2.2 Impedance Measured Through a Transmission Line

Equation (2.17) gives the input impedance for a transmission line of length d, wave impedance Z_o, space frequency β, terminated in load Z_L. Solving (2.17) for Z_L gives

$$Z_L = Z_o \frac{Z_{\text{in}} - jZ_o \tan(\beta d)}{Z_o - jZ_{\text{in}} \tan(\beta d)} \tag{9.5}$$

Most network analyzers, and some amateur antenna analyzers, have this equation built in, so that you don't use it directly, but go through a calibration procedure so the instrument knows the transmission line parameters, especially its electrical length. If you are using an instrument that doesn't do it for you, proceed as follows. Terminating the line in either a short or open circuit will produce a standing-wave pattern on the line with rather broad maxima and sharp nulls. Changing the test frequency has a similar behavior—you will find rather broad maxima of impedance and fairly sharp minima. We are interested in the input impedance nulls because their frequency is more precisely determinable. The lowest frequency for a null is with an open termination and a quarter-wave voltage pattern. Call this frequency f_1 and its free-space wavelength λ_1. Then the equivalent free-space length of the line is:

$$d_e = \lambda_1 / 4 \tag{9.6}$$

With a short-circuit termination, the lowest-frequency pattern for an input null is a half wave. This should occur at $f_2 = 2f_1$. If this relation isn't satisfied to the desired degree, but isn't way out ($f_2 = 3f_1/2$, for example) use the short-circuit result as likely to be the more accurate. The location of an open-circuit termination is less well-defined than a good short, especially a completely closed termination. Another point is that the instrument resolution for impedance may be quite limited. In that case, it is more accurate to note the two frequencies on either side of the null where the readings just change up by one digit, and take their average as the null frequency.

Once you've settled on a value for d_e, use:

$$\beta d = \frac{2\pi d_e}{\lambda} = \frac{2\pi f d_e}{300} = \frac{\pi f d_e}{150} \tag{9.7}$$

in (9.5). Remember that a series resonance gives a minimum impedance magnitude and that the reactance slope is positive, while a parallel resonance gives a peak in impedance magnitude and a negative reactance slope in its immediate neighborhood.

If you are using an instrument that reads S = SWR and $|Z|$, these can be converted into R and $|X|$ as follows. Let Γ be the reflection coefficient. Then, from (2.15), $|\Gamma| = (S - 1)/(S + 1)$ is known. Let $M = |\Gamma|^2$. Then, $M = [(R - Z_o)^2 + X^2]/[(R + Z_o)^2 + X^2]$ which can be solved for R using

$$X^2 = |Z|^2 - R^2 \tag{9.8}$$

$$R = \frac{|Z|^2 + Z_o^2}{2Z_o} \cdot \frac{1 - M}{1 + M} \tag{9.9}$$

9.3 Ranges and Test Enclosures

The purpose of a test range is to put the DUT in an environment as similar to either free space or a free half-space as possible. The transmitted wave should look to the DUT like a plane wave from a single source. Figure 9.2 only illustrates some of the problems in achieving this situation. In many ranges, a very directional antenna is used for the transmission. This has the advantage that relatively little power is reflected from the range bed, and also little goes back to be reflected from the transmitter building. The receiver building should also be out of sight of the antennas and have low reflectivity. Another measure that is sometimes used is to make the range bed slope downhill from the transmit to the receive antenna. If the angles are arranged correctly, the reflected wave will have a parallel direction with the direct wave.

[4] describes a range which simulates a half-space by providing a steel range bed, 30 m by 60 m. Their lowest test frequency was 20 MHz at which the bed is 2λ by 4λ. They spaced the antennas only 10 m apart. The idea in their testing was to use two dipoles and predict the response using analytical methods and NEC2, assuming an infinite ground plane. The measured and NEC2 results agreed generally within ±0.2 dB. Their lowest-frequency dipole was a half-wave at 60 MHz, and they ran it from 20 to 120 MHz. The spacing isn't really far-field even at 60 MHz, but NEC2 would take this into account. This is one of three examples I've read lately that use numerical simulation to validate measurements. Agreement is good, but if there's disagreement the problem may be on either or both sides.

Where is the far-field, and what is not far? Equations (2.3) and (2.4) give the far-field wave expressions for a current element, and, as explained in Chapter 2, these expressions describe a spherical wave front whose amplitudes decrease as $1/r$. A more complete set of expressions usually given in EM texts includes radial as well as transverse field components. The transverse components have terms in $1/(\beta r)$, $j/(\beta r)^2$, and $1/(\beta r)^3$. The cross-product of $\overline{E}$ and $\overline{H}^*$ contains reactive power components as well as traveling-wave components. Traditionally, antenna engineers have taken the far-zone near boundary as $2D/\lambda^2$, where D is the largest dimension of the antenna. This boundary is based on highly directive antennas and can be traced back to Lord Rayleigh's work in 1895 on pin-hole cameras.

Laybros and Combes [5] have examined this question for simple antennas ranging in size from the infinitesimal dipole or loop up to the 1λ dipole. They found that the space around a simple antenna can be divided into three regions: (1) the very-near field in which most of the reactive power is contained, (2) an intermediate range in which the wave is traveling out but is not yet fully spherical, and (3) the far field in which the wave is spherical. The most restrictive condition defining the far-field zone is the ratio of the radial field component to the transverse field component. For –30 dB this is 10λ for short dipoles. In the very-near-field zone, the power flow vector (Poynting vector) has radial and angular components. The angular component is completely reactive and the radial component has real and reactive parts. [5] defines the boundary of this zone as the distance at which the ratio of the reactive to real parts of the radial power density (outward-flowing rather than circulating) is –30 dB. On this basis the boundary is 1.6λ for short dipoles and small loops. This leaves the intermediate zone between 1.6λ and 10λ, quite a spread.

Boswell et. al., in [6], described an effort to measure the radiation efficiency of a 1-m diameter loop at frequencies from 3.6 to 10.1 MHz. This was done with the DUT as the transmitter and a field-strength meter with a calibrated loop as the receiver. Both antennas were kept 1.5 m above a field, and the separations ranged from 20 m to 80 m. They calculated that the discrepancy between far-field theory and their measurements would be 1.1 dB at 20 m and 0.3 dB at 40 m for 3.6 MHz. To arrive at the antenna's efficiency, they used the measured field strength and a theoretical calculation for a loss-free loop in the same environment. While this procedure may seem to lack experimental rigor, a quarter-wave monopole at 3.6 MHz is about 20 m tall so that measurement by comparison to a low-loss reference antenna is generally not practical.

At frequencies above HF, various kinds of enclosures are used. The most common one is the anechoic chamber. The anechoic chamber is a room whose interior surfaces are covered by EM field-absorbing cones. Besides using them as electrically large test chambers, a lot of science has been developed over the last 20 years to use near-field measurements to predict the far-field behavior of an antenna.

9.4 The Wheeler Cap and Variations

For many small antennas, the most important parameter to measure is efficiency. A common method to do this is by substituting a calibrated or well-understood dipole for the small antenna and measure the difference in power transferred through the system. Since conduction loss doesn't scale with frequency, one can't scale the antennas to a convenient size, so that testing at the desired operating frequency is required and possibly inconvenient. In 1959, Wheeler [7] published a paper that proposed that a conducting shell enclosure be used to measure the efficiency of a small antenna. The shell is a sphere whose radius is r such that $2\pi r/\lambda = 1$ radian. The size of the shell is therefore called by Wheeler a radiansphere, and others have called it the *Wheeler Cap* ever since. Its purpose is to block the small antenna from radiating, so that any power absorbed is loss power. A direct measurement of the input impedance would then give the loss resistance. Testing the same antenna in an unrestricted space or half-space would give the sum of radiation and loss resistances, so that the two measurements together would allow finding the radiation resistance and efficiency. The procedure assumes that the current distribution is the same with and without the shell. In his paper, Wheeler says that the size and shape of the shielding shell is not critical; the sphere is used for theoretical convenience. He doesn't give any detailed experimental results.

Glenn Smith [8] published an analysis of the Wheeler Cap based on a theoretical solution for the currents in a loop and the shield. As this is a thought and calculation experiment, it represents the best accuracy that can be expected for this combination. If we call the radiation resistance R_R, the loss resistance R_L, and the resistance measured without the cap R_o, then the efficiency is:

$$E = \frac{R_o - R_L}{R_o} = \frac{R_R}{R_o} \qquad (9.10)$$

This assumes a series model for the antenna and its impedance, and is the expression Smith evaluated. He numerically tested the method for loop circumference from 0.05λ to 0.9λ with both loop and shield made of copper. Below 0.4λ, the method was accurate to three figures. Above this point there were large differences between the loss resistance with the shield and without the shield, especially at the parallel-resonant size. Smith attributes this difference to a change in the current distribution at resonance, although I don't see a difference in shape in his plots, just a scale difference.

People working on microstrip antennas, which are resonant and not electrically small, have been using the radiation-shield method since the '80s [9–11]. Pozar and Kaufman [9] found the method to be more accurate and precise than direct gain measurements. Johnston and McRory [10] proposed a method in which a section of waveguide with sliding shorts is used as the shield cavity. Instead of measuring input impedance or power transfer, they measure the S parameters, and use an efficiency expression based on S parameters. They did both a transmission (two antennas in the cavity) test and a reflection test, and found the reflection method to be more precise, and more precise than the cap results reported in [9]. In the reflection method, one short is held stationary and the other is moved to various positions over a quarter guide wavelength. At each position $S_{11\text{wgs}}$, the reflection coefficient, is measured. These values will lie on a circle in the complex plane. The minimum and maximum distance from this circle to $S_{11\text{fs}}$, the free-space reflection coefficient, determines the antenna's efficiency. This method requires a more elaborate construction and more data points per frequency than the simple cap method, but may be worthwhile for the improved precision.

H. Choo et. al. [11] used a rectangular box as the shield cavity, and it was not electrically small. However, they got good results by keeping the cavity-resonance mode frequencies away from the test frequencies. They also found that using (9.10), the resistance-based expression for efficiency, gave poor results near the antenna's resonant frequency, which is also the desired operating frequency. They found that modeling the antenna near resonance as a parallel RLC circuit requires writing the efficiency using conductances, and this worked well. Using the same subscript meanings as for (9.10), the conductance-based efficiency is:

$$E = \frac{G_o - G_L}{G_o} = \frac{G_R}{G_o} \tag{9.11}$$

9.4.1 Series and Parallel Effects

The results of [11] prompt the questions "What's the appropriate loss model?" and "What happens when you combine simple series and parallel models into series and parallel-resonant circuits?" Figure 9.4(a) shows the equivalent circuit for a loop below its resonance. By its physics it is most appropriate to model it as a series RL, and the power absorbed is most simply expressed as I^2R_1. A capacitor-loaded dipole with a lossy dielectric can be modeled as a parallel RC circuit and the power absorbed by the conductance is most simply expressed as $V^2G_2 = V^2/R_2$. In the first case, an increase in R causes an increase in power absorbed, in the second case an increase in R causes a decrease in power absorbed.

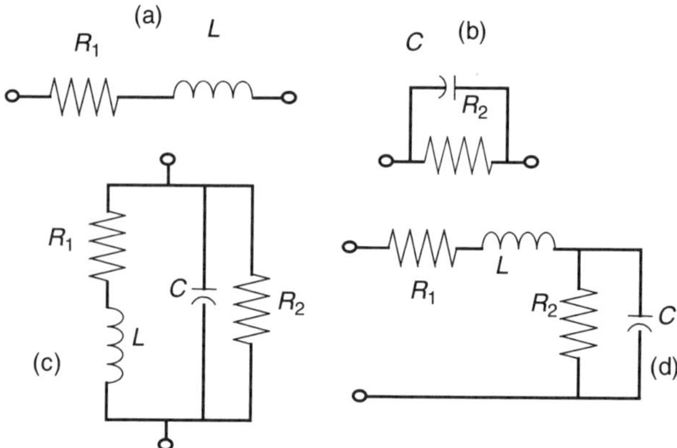

Figure 9.4: (a) Small loop equivalent circuit. (b) Dielectric-loaded dipole equivalent circuit. (c) Parallel-resonant circuit. (d) Series-resonant circuit.

In Figure 9.4(c), the input admittance is:

$$\begin{aligned} Y &= G_2 + j\omega C + \frac{R_1 - j\omega L}{R_1^2 + \omega^2 L^2} \\ &\approx G_2 + \frac{R_1}{\omega^2 L^2} + j\left(\omega C - \frac{1}{\omega L}\right), \quad \omega L >> R_1 \end{aligned} \tag{9.12}$$

Observe that, as long as the inequality holds, an increase in either R_1 or G_2 causes the input conductance to increase. Likewise, the input impedance for the series-resonant circuit in Figure 9.4(d) is:

$$\begin{aligned} Z &= R_1 + j\omega L + \frac{G_2 - j\omega C}{G_2^2 + \omega^2 C^2} \\ &\approx R_1 + \frac{G_2}{\omega^2 C^2} + j\left(\omega L - \frac{1}{\omega C}\right), \quad \omega C >> G_2 \end{aligned} \tag{9.13}$$

Now, as long as the inequality holds, an increase in either R_1 or G_2 will cause the input resistance to increase. We may conclude from these results that the kind of resonance is as important as what the loss mechanism does to the component resistances.

Example 9.2 Efficiency for Two Hypothetical Antennas Connected in Resonant Circuits

I decided to do a wideband numerical test of the circuits in Figure 9.4(c) and (d). I wanted to get resonance at $\omega = 1$ and reasonably high Q, so I chose

$R_{R1} = R_{L1} = 0.5\Omega$, $G_{R2} = G_{L2} = 0.005$ S, $L = 10$, and $C = 0.1$. This is a totally hypothetical situation because the resistances are not functions of frequency. The efficiency should be 0.5 at all frequencies. Using MATLAB, I calculated the input impedance and admittance for each circuit over a frequency range 0.001 to 2. As you can see in Figures 9.5 and 9.6, neither the conductance-based nor resistance-based version of efficiency gets the right answer over the whole range. The higher value is always closer to correct. (9.10) is appropriate for the series-resonance region and case, (9.1) is appropriate for the parallel-resonance region and case. I leave the low-frequency behavior for you to explain in the problems.

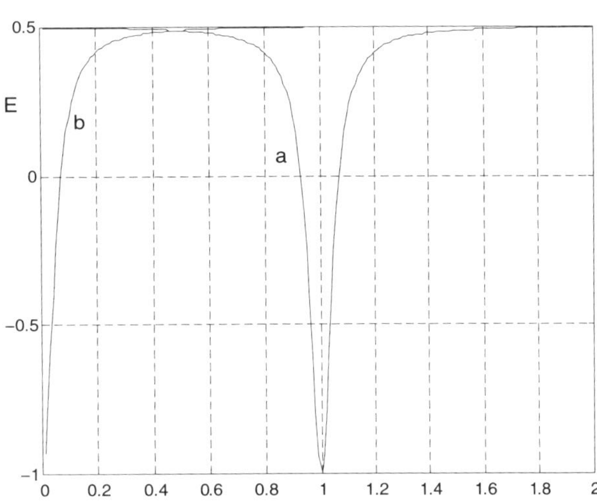

Figure 9.5: Efficiency for a parallel-resonant circuit. (a) Resistance-based. (b) Conductance-based.

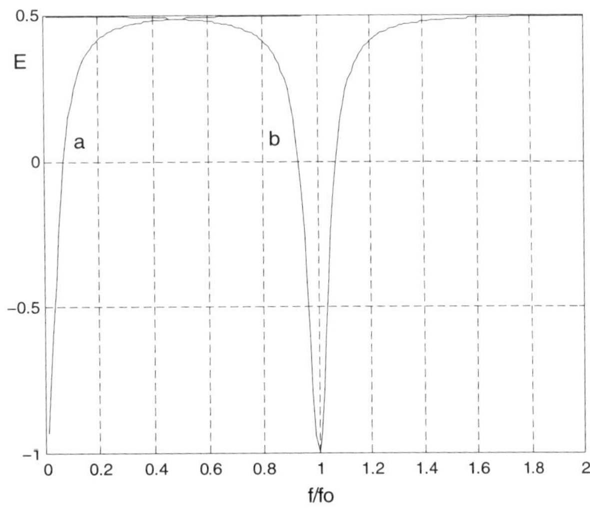

Figure 9.6: Efficiency for a series resonant circuit. (a) Resistance-based. (b) Conductance-based.

References

[1] J. R. Hallas, "A Look at Some High-End Antenna Analyzers," Product Review in *QST*, vol. 89, no. 5, pp. 65–69, May 2005.

[2] *The ARRL Handbook for Radio Amateurs*, 71st ed., ARRL, Newington, CT, 1994.

[3] D. B. Miron, "The HF Short Fat Dipole: A Progress Report," antennex.com, September 2004.

[4] M. J. Alexander and M. J. Salter, "Design of Dipole and Monopole Antennas with Low Uncertainties," *IEEE Trans. Instr. and Meas.*, vol. 46, no. 2, pp. 539–543, April 1997.

[5] S. Laybros and P. F. Combes, "On Radiation-Zone Boundaries of Short, λ/2, and λ Dipoles," *IEEE Antennas and Propagation Magazine*, vol. 46, no. 5, pp. 53–64, October 2004.

[6] Alan Boswell, et. al., "Performance of a Small Loop Antenna in the 3-10 MHz Band," *IEEE Antennas and Propagation Magazine*, vol. 47, no. 2, pp. 51–57, April 2005.

[7] H. A. Wheeler, "The Radiansphere Around a Small Antenna," Proc. IRE, vol. 47, no. 8, pp. 1325–1331, August 1959.

[8] G. S. Smith, "An Analysis of the Wheeler Method of Measuring the Radiating Efficiency of Antennas," *IEEE Trans. Ant. and Prop.*, vol. 25, no. 4, pp. 552–556, July 1977.

[9] D. M. Pozar and B. Kaufman, "Comparison of Three Methods of Printed Antenna Efficiency," *IEEE Trans. Ant. and Prop.*, vol. 36, no. 1, pp. 136–139, January 1988.

[10] R. H. Johnston and J. G. McRory, "An Improved Small Antenna Radiation-Efficiency Measurement Method," *IEEE Antennas and Propagation Magazine*, vol. 40, no. 5, pp. 40–48, October 1998.

[11] H. Choo et. al., "On the Wheeler Cap Measurement of the Efficiency of Microstrip Antennas," *IEEE Trans. Ant. and Prop.*, vol. 53, no. 7, pp. 2328–2332, July 2005.

Chapter 9 Problems

Section 9.2.1

9.1 Prove the assertion that, for a load modeled as $R + jX$, the minimum SWR occurs at $X = 0$.

9.2 Prove that, for a load modeled as $G + jB$, minimum SWR occurs at $B = 0$.

9.3 Prove (9.1), the expression for B that minimizes the SWR for a fixed $R + jX$ load.

9.4 Derive (9.3) and (9.4), expressions that determine the unknown load given B and S for minimum SWR.

9.5 In Example 9.1, given that the shunt $C = 1712$ pF for SWR = 1 at 3.95 MHz, what were R, X, and the terminal inductance?

9.6 In Example 9.1, given that SWR = 1.7 at 3.92 and 3.995 MHz, (a) what is the SWR = 2 bandwidth? (b) What is the half-power bandwidth?

9.7 For a load modeled as $G + jB$, find the series X that minimizes SWR. What is this minimum SWR?

9.8 Having solved 9.7, given experimental values for the X that minimizes SWR, and the minimum SWR, find expressions for G and B.

Section 9.2.2

9.9 Derive (9.5), the expression for Z_L given Z_{in}.

9.10 Derive an equivalent to (9.5) using admittances.

9.11 For a series RLC circuit, derive an expression for the reactance slope at resonance.

9.12 For a parallel RLC circuit, derive an expression for the reactance slope at resonance.

9.13 Derive (9.9), the expression for R given $|Z|$ and SWR.

Section 9.3

9.14 Consider a slant range design problem. The minimum vertical distance for the transmit antenna is 2 m and the receive antenna is on the same horizontal line as the transmit antenna. The distance between the antennas is 100 m. What should the slant angle of the range bed be in order that all rays in the transmit antenna's half-power vertical beamwidth that are reflected from the range bed pass below the receive antenna? You may assume that the angles of reflection and incidence are equal.

9.15 Consider the steel bed range described in the text and in [3]. Assume two 2.5-m vertical dipoles centered 2 m above the bed and placed with horizontal symmetry. Model the transmit dipole as being fed from a 5W 75Ω source. Put a 75Ω resistor in the center segment of the receive dipole. Use NEC2 to find the receive voltage over an infinite ground plane from 20 to 120 MHz. Then simulate the range bed with a wire grid and a free-space environment and repeat the calculation. Comment on the results.

9.16 The loop described in [5] used a smaller loop as a drive element and a series capacitor at the top of the main loop as a tuning element. Model this arrangement in NEC2, tune it up as described in [5], including the ground constants and conductor loss, and see if you can duplicate their field-strength measurements.

Section 9.4

9.17 Use one of the cylinder models described in Chapter 6 to model a Wheeler-Cap over a PEC ground. Model a 0.05λ radius half-loop inside it. Use NEC2 to find the input impedance and the E-field at 10λ. Then model the loop without the shield and find the same variables. What is the ratio of the **E** fields in dB? Find the efficiencies from (9.10) and (9.11) and compare them to the structure efficiency reported by NEC.

Section 9.4.1

9.18 At resonance, what is the relationship, including phase, for currents in the two resistors in Figure 9.4(c)?

9.19 What should be the relation between the resistors in Figure 9.4(d) so that they absorb equal powers at resonance?

9.20 Explain the DC behavior of the efficiencies calculated in Example 9.2.

APPENDIX A

The Mathematics of Antenna Orientation

The field expressions for the vertical dipole and horizontal loop given in Chapter 2 are simple because they happen to be in the best orientation in the spherical coordinate system for the purpose. For practical reasons, it is useful to have field expressions for other orientations in the same coordinate system. Specifically, in this appendix I show how to find the expressions for a horizontal dipole on the x axis and a vertical loop in the x-z plane. The method is to express the starting expression in terms of (x,y,z) coordinates, relabel the axes to simulate rotation of the antenna, and then convert the result back to angle coordinates. To do this, we need to express unit vectors from one system in terms of those of another. This is shown in the next section.

A.1 Unit-Vector and Coordinate Variable Relations

Figure 2.1 in Chapter 2 shows spherical (r,θ,ϕ), cylindrical (ρ,ϕ,z), and rectangular (x,y,z) coordinate systems. To get the relations we need, I take sections of this figure as shown in Figure A.1. Figure A.1(a) shows the vertical plane at ϕ, the (ρ,z) plane. ρ is the projection of r into the (x,y) plane. The variable relations we need from this figure are:

$$\sin(\theta)=\frac{\rho}{r},\cos(\theta)=\frac{z}{r},r^2=\rho^2+z^2 \quad \text{(A.1)}$$

We need to express the angle unit vectors in terms of the rectangular unit vectors. Figure A.1(b) shows the unit vectors and a triangle for which $\hat{\theta}$ is the hypotenuse. Since, by definition, the length of $\hat{\theta}$ is 1, the other side lengths are cos(θ) and sin(θ). As vectors, each of these sides is the length times the unit vector going its way. The vector sum is:

$$\hat{\theta} = \hat{\rho}\cos(\theta) - \hat{z}\sin(\theta) \quad \textbf{(A.2)}$$

ρ and ϕ are in the (x,y) plane, so we go next to Figure A.1(c). The relations we need from this figure are:

$$\cos(\phi) = \frac{x}{\rho}, \sin(\phi) = \frac{y}{\rho}, \rho^2 = x^2 + y^2 \quad \textbf{(A.3)}$$

From the unit-vector triangle in Figure A.1(d)

$$\hat{\rho} = \hat{x}\cos(\phi) + \hat{y}\sin(\phi) \quad \textbf{(A.4)}$$

Finally, from Figure A.1(e) we have

$$\hat{\phi} = -\hat{x}\sin(\phi) + \hat{y}\cos(\phi) \quad \textbf{(A.5)}$$

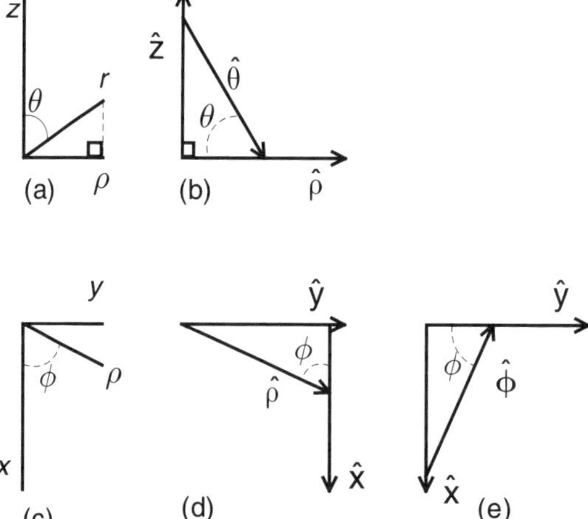

Figure A.1: Sketches to show the relations between spherical, cylindrical, and rectangular variables.

We now have equations to replace the spherical variables with the rectangular ones. We also need some help going from (x,y) vectors back to angle vectors. Figure A.2 shows two sketches to help us. From Figure A.2(a):

$$\hat{x} = \hat{\rho}\cos(\phi) - \hat{\phi}\sin(\phi) \quad \textbf{(A.6)}$$

and from Figure A.2(b)

$$\hat{y} = \hat{\phi}\cos(\phi) + \hat{\rho}\sin(\phi) \quad \textbf{(A.7)}$$

Figure A.2: Sketches to get rectangular unit vectors in terms of cylindrical ones.

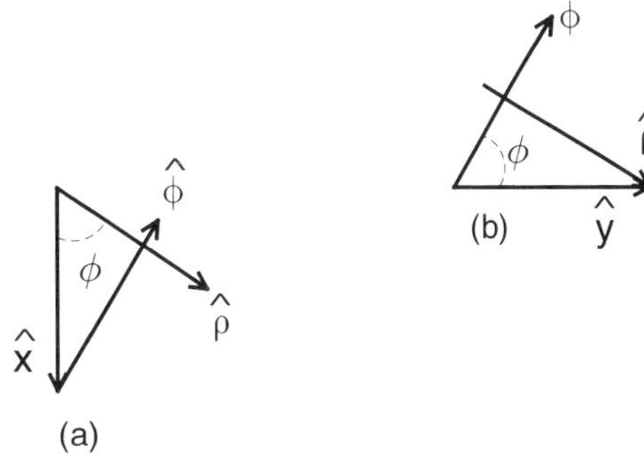

A.2 The Horizontal Dipole

In this and the following section, I only deal with that part of the field expression which depends on angles. For the vertical dipole this is:

$$D = \sin(\theta)\hat{\theta} \tag{A.8}$$

From the previous section, first replace $\hat{\theta}$ then $\hat{\rho}$.

$$\begin{aligned} D &= \sin(\theta)\left(\hat{\rho}\cos(\theta) - \hat{z}\sin(\theta)\right) \\ &= \sin(\theta)\left((\hat{x}\cos(\phi) + \hat{y}\sin(\phi))\cos(\theta) - \hat{z}\sin(\theta)\right) \\ &= \hat{x}\cos(\phi)\sin(\theta)\cos(\theta) + \hat{y}\sin(\phi)\sin(\theta)\cos(\theta) - \hat{z}\sin^2(\theta) \\ &= \hat{x}\frac{xz}{r^2} + \hat{y}\frac{yz}{r^2} - \hat{z}\frac{x^2+y^2}{r^2} \end{aligned} \tag{A.9}$$

Now, to move the dipole from the z axis to the x axis, I relabel the z axis to x. This is a right-handed coordinate system, so $z \to x$, $x \to y$, $y \to z$. This leads to

$$D = \frac{y^2+z^2}{r^2}\hat{x} + \frac{xy}{r^2}\hat{y} + \frac{zx}{r^2}\hat{z} \tag{A.10}$$

This is D for the dipole centered on the x axis. To get back to an expression in angles, it is best to rearrange (A.10), as much as possible, so it has factors that look like the equations in section A.1 for the angular unit vectors. Full direct substitution leads to long expressions that raise the probability of mistakes.

$$\begin{aligned} D &= \frac{y}{r}\left(-\frac{y}{r}\hat{x} + \frac{x}{r}\hat{y}\right) - \frac{z^2}{r^2}\hat{x} + \frac{zx}{r^2}\hat{z} \\ &= \sin(\phi)\sin^2(\theta)\left(-\sin(\phi)\hat{x} + \cos(\phi)\hat{y}\right) - \cos^2(\theta)\hat{x} + \cos(\theta)\sin(\theta)\cos(\phi)\hat{z} \\ &= \hat{\phi}\sin(\phi)\sin^2(\theta) - \cos^2(\theta)\hat{x} + \cos(\theta)\sin(\theta)\cos(\phi)\hat{z} \end{aligned} \tag{A.11}$$

Now using (A.6)

$$\begin{aligned} D &= \hat{\phi}\sin(\phi)\sin^2(\theta) - \cos^2(\theta)\left(\hat{\rho}\cos(\phi) - \hat{\phi}\sin(\phi)\right) + \cos(\theta)\sin(\theta)\cos(\phi)\hat{z} \\ &= \hat{\phi}\sin(\phi) + \cos(\phi)\cos(\theta)\left(-\cos(\theta)\hat{\rho} + \sin(\theta)\hat{z}\right) \\ &= \hat{\phi}\sin(\phi) - \hat{\theta}\cos(\phi)\cos(\theta) \end{aligned} \quad \textbf{(A.12)}$$

Does this make physical sense? When the dipole was on the z axis, the reference direction for $\overline{\mathrm{E}}$ was $\hat{\theta}$, which is generally opposite to the current. In the present case the current is along the positive direction of the x axis. For $\phi = 0$ (A.12) becomes $D = -\hat{\theta}\cos(\theta)$. This is in the x-z plane. In the upper half, $\hat{\theta}$ is in the same direction as the current, $\cos(\theta) > 0$, so the result is right. In the lower half of the x-z plane, $\hat{\theta}$ is opposite to the current and $\cos(\theta) < 0$, so the result is right again. Also the amplitude variation is correct. For $\phi = \pi/2$, (A.12) becomes $D = \hat{\phi}$ which again has the correct direction and amplitude variation.

Following the same procedure for a dipole moving from the z to the y axis, $z \rightarrow y$, $y \rightarrow x$, $x \rightarrow z$, I found:

$$D = -\hat{\phi}\cos(\phi) - \hat{\theta}\cos(\theta)\sin(\phi) \quad \textbf{(A.13)}$$

A.3 The Vertical Loop

For a horizontal loop centered in the x-z plane, the angle-dependent part of the field expression is

$$D = \hat{\phi}\sin(\theta) \quad \textbf{(A.14)}$$

Using the equations in section A.1 this converts to

$$D = -\hat{x}\frac{y}{r} + \hat{y}\frac{x}{r} \quad \textbf{(A.15)}$$

Moving the loop to the x-z plane is equivalent to moving its axle from the z to the y axis. This is a left-handed rotation, $z \rightarrow y$, $y \rightarrow x$, $x \rightarrow z$.

$$\begin{aligned} D &= \hat{x}\frac{z}{r} - \hat{z}\frac{x}{r} \\ &= \cos(\theta)\left(\hat{\rho}\cos(\phi) - \hat{\phi}\sin(\phi)\right) - \hat{z}\cos(\phi)\sin(\theta) \\ &= \hat{\theta}\cos(\phi) - \hat{\phi}\sin(\phi)\cos(\theta) \end{aligned} \quad \textbf{(A.16)}$$

Moving the loop from horizontal to the y-z plane moves the loop's axle from the z to the x axis. Following the same procedures, I found:

$$D = -\hat{\theta}\sin(\phi) - \hat{\phi}\cos(\phi)\cos(\theta) \tag{A.17}$$

Appendix A Problems

A.1 Derive equation (A.13).

A.2 Derive equation (A.17).

APPENDIX B

The Parallel-Ray Approximation

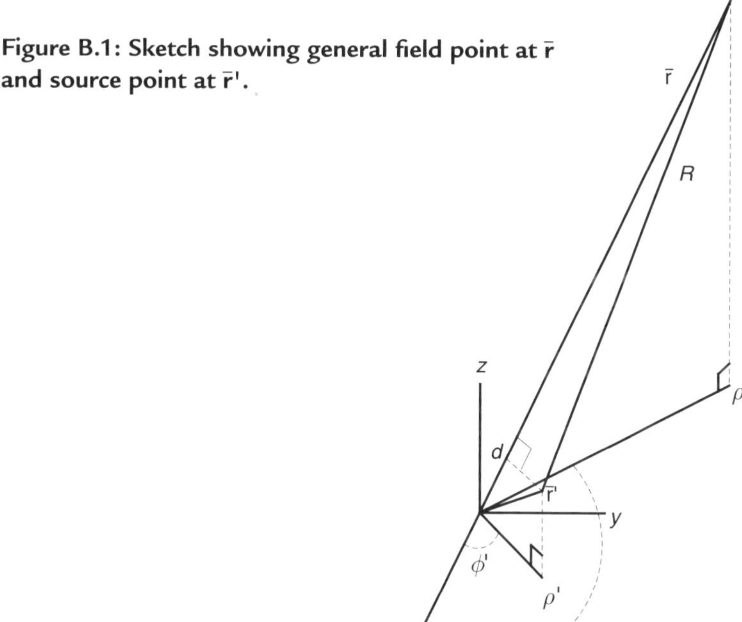

Figure B.1: Sketch showing general field point at r̄ and source point at r̄'.

In Chapter 2, equations (2.3) and (2.4) give the electromagnetic fields due to a source on the z axis at the origin of coordinates. In Appendix A, we find what happens for various re-orientations of the source. In this Appendix, we find how to deal with a source moved away from the origin. Figure B.1 shows a field point at $\bar{r}$ and a source point at $\bar{r}'$. The distance between them is the line R. The propagation part of the field expression is:

$$f(R) = \frac{e^{-j\beta R}}{R}, \; R = |\bar{r} - \bar{r}'| \quad \textbf{(B.1)}$$

The parallel-ray approximation is that the field point is so far away from both the origin and the source point that the lines for r and R are essentially parallel. This leads to:

$$R = r - d \tag{B.2}$$

where d is the length of the projection of $\bar{r}'$ on $\bar{r}$. This is the vector dot product of $\bar{r}'$ with the unit vector for $\bar{r}$.

$$\begin{aligned} d &= \bar{r}' \cdot \hat{r} = (\hat{\rho}' + \hat{z}') \bullet (\hat{\rho} \sin(\theta) + \hat{z} \cos(\theta)) \\ &= (\rho' (\hat{x} \cos(\phi') + \hat{y} \sin(\phi')) + \hat{z} z') \bullet (\sin(\theta)(\hat{x} \cos(\phi) + \hat{y} \sin(\phi)) + \hat{z} \cos(\theta)) \\ &= \rho' \sin(\theta) \cos(\phi' - \phi) + z' \cos(\theta) \end{aligned} \tag{B.3}$$

Now replacing R with $r - d$ in (B.1),

$$f(R) = \frac{e^{-j\beta(r-d)}}{r-d} = \frac{e^{-j\beta(r-d)}}{r\left(1 - \frac{d}{r}\right)} = \frac{e^{-j\beta r}}{r}\left(1 + \frac{d}{r}\right) e^{j\beta d} \tag{B.4}$$

The result has the propagation function based at the origin, an amplitude-shift factor, and a phase-shift factor. Because d is tiny compared to r, usually the d/r term is dropped. However, βd may be a significant angle and so the phase term is kept in cases where this is true or where it makes the small difference between large subtracted terms.

Appendix B Problems

B.1 Suppose there are two vertical current sources on the y axis at $+a$ and $-a$. Find the simplest possible expression for the combined electric field if (a) the currents have the same phase, and (b) the currents have opposite phases. You will need equation (2.3) and results from Appendix A.

B.2 Repeat problem B.1 except that the sources are on the x axis.

B.3 Repeat problem B.1 except that the sources are horizontal along the y axis.

B.4 Repeat problem B.1 except that the sources are parallel to the x axis.

APPENDIX C

The Small Loop

Loop antennas come in a variety of shapes but the most common single-turn shapes are the circle and the rectangle. In this appendix I use results from Appendices A and B to find the radiated field for a small circular loop carrying a uniform (independent of position) current. Figure C.1(a) shows the setup for a loop in the *x-y* plane, and Figure C.1(b) shows the decomposition of a current-length element. As pointed out in Chapter 2, the vector orientation of the radiated fields depends on the direction of the current source. In electromagnetic theory, the current density is given a vector character, and when current density times volume, $\bar{\text{J}}dV$, is converted into total current times length, it is the length element that is assigned the vector property to show the direction of current flow as $Id\bar{\text{L}}$. In the present case, the current flows around the loop in the ϕ' direction. At each point on the loop, the element of length is $ad\phi'$. From Figure C.1(b), we see that this corresponds to a *y*-directed component of length $dy' = a\cos(\phi')d\phi'$ and a $-x$-directed component$dx' = a\sin(\phi')d\phi'$. From equation (2.3) and Appendix A, if these current length elements were at the origin, their radiated fields would be:

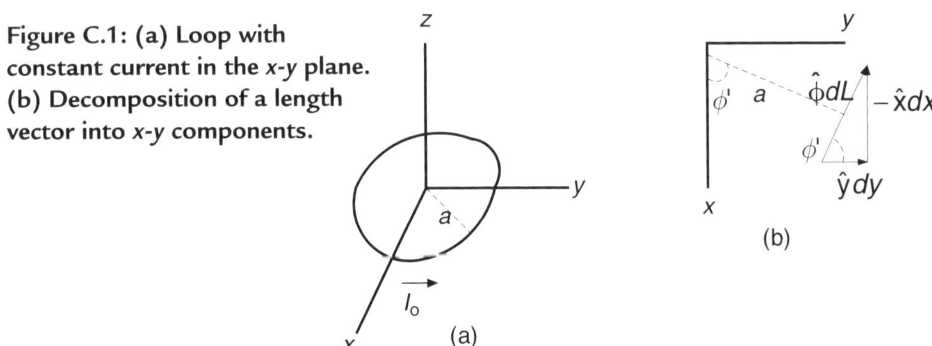

Figure C.1: (a) Loop with constant current in the *x-y* plane. (b) Decomposition of a length vector into *x-y* components.

from the x component,

$$\bar{\mathrm{E}} = -\frac{j\eta}{2\lambda} I_o a \sin(\phi') d\phi' \left(\hat{\phi} \sin(\phi) - \hat{\theta} \cos(\phi) \cos(\theta)\right) \frac{e^{-j\beta r}}{r} \tag{C.1}$$

and from the y-component,

$$\bar{\mathrm{E}} = -\frac{j\eta}{2\lambda} I_o a \cos(\phi') d\phi' \left(\hat{\phi} \cos(\phi) + \hat{\theta} \sin(\phi) \cos(\theta)\right) \frac{e^{-j\beta r}}{r} \tag{C.2}$$

Their fields combine to make:

$$\bar{\mathrm{E}} = -\frac{j\eta}{2\lambda} I_o a d\phi' \left(\hat{\phi} \cos(\phi' - \phi) - \hat{\theta} \sin(\phi' - \phi) \cos(\theta)\right) \frac{e^{-j\beta r}}{r} \tag{C.3}$$

However, they are not at the origin but are closer to the field point by a distance d which is, from Appendix B,

$$d = a \cos(\phi' - \phi) \sin(\theta) \tag{C.4}$$

This causes a phase advance (equivalent to being nearer in time) of βd. The net field for our current element is now:

$$\bar{\mathrm{E}} = -\frac{j\eta}{2\lambda} I_o a d\phi' \left(\hat{\phi} \cos(\phi' - \phi) - \hat{\theta} \sin(\phi' - \phi) \cos(\theta)\right) \frac{e^{-j\beta r}}{r} e^{j\beta a \cos(\phi' - \phi) \sin(\theta)} \tag{C.5}$$

At this point we could sum the fields from all the current-length elements going around the loop, But we can save a little work by using symmetry. At each point on the loop, there is a current-length element diametrically across the loop pointing in the opposite direction. Its distance displacement from the field point will be opposite (negative of) as well. Using these facts, the total field from a pair of such elements is:

$$\bar{\mathrm{E}} = \frac{\eta}{\lambda} I_o a d\phi' \left(\hat{\phi} \cos(\phi' - \phi) - \hat{\theta} \sin(\phi' - \phi) \cos(\theta)\right) \frac{e^{-j\beta r}}{r} \sin\left(\beta a \cos(\phi' - \phi) \sin(\theta)\right) \tag{C.6}$$

Because the loop is electrically small we can replace $\sin(\beta a \ldots)$ with its argument. Now we are ready to write the total field, bearing in mind that our last expression is for element pairs so we only need to integrate from 0 to π.

$$
\begin{aligned}
\overline{\mathrm{E}} &= \frac{\eta}{\lambda} I_o a \frac{e^{-j\beta r}}{r} \int_0^{\pi} d\phi' \beta a \sin(\theta) \cos(\phi' - \phi) \left(\hat{\phi} \cos(\phi' - \phi) - \hat{\theta} \sin(\phi' - \phi) \cos(\theta) \right) \\
&= \frac{\eta}{\lambda} I_o a \frac{e^{-j\beta r}}{r} \int_0^{\pi} d\phi' \beta a \sin(\theta) \cos(\phi') \left(\hat{\phi} \cos(\phi') - \hat{\theta} \sin(\phi') \cos(\theta) \right) \\
&= \frac{\eta}{\lambda} I_o \beta a^2 \sin(\theta) \frac{e^{-j\beta r}}{r} \int_0^{\pi} d\phi' \left(\hat{\phi} \cos^2(\phi') - \hat{\theta} \frac{\sin(2\phi')}{2} \cos(\theta) \right) \\
&= \frac{\eta}{2\lambda} I_o \beta \pi a^2 \sin(\theta) \hat{\phi} \frac{e^{-j\beta r}}{r} \\
&= \eta \pi \frac{A}{\lambda^2} I_o \sin(\theta) \hat{\phi} \frac{e^{-j\beta r}}{r}
\end{aligned}
\tag{C.7}
$$

In going from the first to the second line, I've used the fact that the antenna looks the same from any ϕ, so nothing is lost by setting $\phi = 0$. In the last line A is the loop area. It turns out that the result is the same for a loop of any shape—the field doesn't depend on the shape, only the enclosed area. This can be proved by calculating the vector magnetic potential for an arbitrary shape.

Unlike the dipole, the loop's field scales by wavelength squared—that is, by measuring the area in square wavelengths. This is due to two effects, the length/wavelength effect we've seen in the dipole and the spacing/wavelength effect that gives a net field from two separated opposing currents. This can be seen more clearly by doing the field calculation for a rectangular loop, as requested in Problem C.1.

Appendix C Problems

C.1 Find the electric field expression for a rectangular loop lying in the x-y plane, centered at the origin. Let its length in the x direction be a and its length in the y direction be b.

C.2 Find the radiated electric field for an equilateral triangle lying in the x-y plane. One side runs from $-a/2$ to $a/2$ on the y axis, and one corner lies on the $+x$ axis. What is the physical reason for the extra phase term that doesn't appear in (C.7)?

C.3 Find the value of a/λ for which βa differs from $\sin(\beta a)$ by 1%.

C.4 Show the steps going from (C.1) and (C.2) to (C.3).

APPENDIX D

The Proximity Effect

D.1 Current Distribution

In the late 1960s Glenn Smith analyzed the problem of what happens to the current distribution in long parallel wires that are close to one another in terms of their radius. His work was published in a Harvard technical report [1], and a paper [2] in 1972. This Appendix reworks the analysis given in the paper, makes some comments, and describes a simpler method to find the coefficients in the Fourier series for the normalized current densities.

D.1.1 Problem Formulation and Reduction to a System of Linear Equations

Figure D.1 shows the layout and notation for the geometry. The plan of attack is to develop the vector magnetic potential near wire m due to all currents, then apply a boundary condition to get the integral equation for the surface current. This requires an integration of the contribution from each element of current density, and that requires writing an expression for the distance from the source point at (a, θ', z') on wire n to the observation point at (r, θ, z) on wire m.

The overall distance is R_{mn}. Let r_{mn} be the distance between the two points projected into an x-y plane such as is occupied by the circles in the figure. $R_{mn}^2 = (z - z')^2 + r_{mn}^2$. r_{mn} in turn has x and y components. The x part is made up of the center-to-center distance plus the projections of a and r, $d_x = (m - n)d + r\cos(\theta) - a\cos(\theta')$, $m > n$, $d_x = -[(m - n)d + r\cos(\theta) - a\cos(\theta')]$, $m < n$.

Figure D.1: Cylindrical conductors of length 2*h*. The transverse view shows half the length, and is not to scale. *D* >> *Nd*, where *N* is the number of wires.

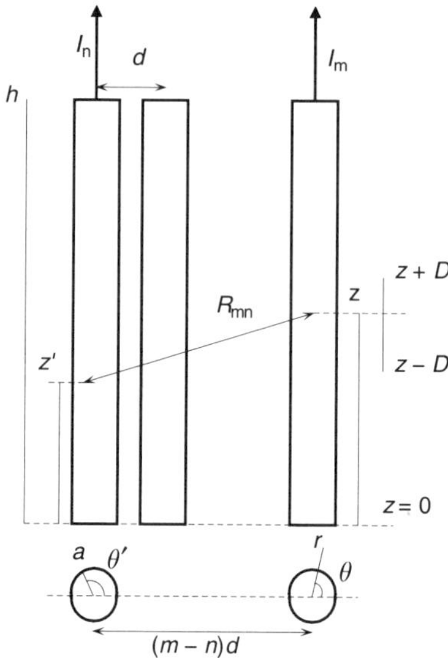

The y component is $d_y = |r\sin(\theta) - a\sin(\theta')|$. The complete expressions are:

$$r_{mn}^2 = \left[(m-n)d + r\cos(\theta) - a\cos(\theta')\right]^2 + \left[r\sin(\theta) - a\sin(\theta')\right]^2 \quad \textbf{(D.1)}$$

$$R_{mn}^2 = (z-z')^2 + \left[(m-n)d + r\cos(\theta) - a\cos(\theta')\right]^2 + \left[r\sin(\theta) - a\sin(\theta')\right]^2 \quad \textbf{(D.2)}$$

The currents are assumed uniform and not a function of z. This requires all dimensions to be much less than the wavelength. The current density and the vector magnetic potential are z-directed. The current density on wire n is

$$K_n(\theta) = \frac{I}{2\pi a} g_n(\theta) = \frac{I}{2\pi a}\left(1 + \sum_{p=1}^{q} a_{np}\cos(p\theta)\right) \quad \textbf{(D.3)}$$

Only a cosine series is needed because the current above the centerline should mirror that below the centerline. With N wires, the vector magnetic potential for wire m is:

$$\begin{aligned} A_m(r,\theta,z) &= \frac{\mu_o}{4\pi}\sum_{n=1}^{N}\int_{-h}^{h}\int_{-\pi}^{\pi} K_n(\theta')\frac{e^{-j\beta R_{mn}}}{R_{mn}} a\,d\theta'\,dz' \\ &= \frac{\mu_o I}{8\pi^2}\sum_{n=1}^{N}\int_{-h}^{h}\int_{-\pi}^{\pi}\left(1 + \sum_{p=1}^{q} a_{np}\cos(p\theta')\right)\frac{e^{-j\beta R_{mn}}}{R_{mn}} d\theta'\,dz' \end{aligned} \quad \textbf{(D.4)}$$

Since we are going to apply a boundary condition at the surface of each wire, we are only interested in r close to a. If z and z' are far apart compared to the width of the group Nd, the angular offsets in the distance function can be ignored. That's the reason for marking the region $z \pm D$ on the figure. For z' outside this space the distance function is:

$$R_{mn}^2 \approx (z-z')^2 + d^2(m-n)^2 \tag{D.5}$$

Using this approximation splits the integral in two parts, one that is a function of $(r,\theta,z$, and one that is only a function of z. Furthermore, since $D << \lambda$ we can approximate the exponential by 1. These ideas lead to:

$$A_m(r,\theta,z) = \frac{\mu_o I}{8\pi^2}\sum_{n=1}^{N}\int_{z-D}^{z+D}\int_{-\pi}^{\pi}\left(1+\sum_{p=1}^{q} a_{np}\cos(p\theta')\right)\frac{1}{R_{mn}}d\theta' dz' + F(z) \tag{D.6}$$

The integral in z' has the form:

$$\int_{-D}^{D}\frac{du}{\sqrt{u^2+r_{mn}^2}} = \ln\left(\frac{D+\sqrt{D^2+r_{mn}^2}}{-D+\sqrt{D^2+r_{mn}^2}}\right) \approx 2\ln(2D) - 2\ln(r_{mn}) \tag{D.7}$$

By themselves, the cosines in the series integrate to zero over the circle.

$$A_m(r,\theta,z) = \frac{\mu_o I}{4\pi^2}\left[2\pi\ln(2D) - \sum_{n=1}^{N}\int_{-\pi}^{\pi}\left(1+\sum_{p=1}^{q}a_{np}\cos(p\theta')\right)\ln(r_{mn})d\theta'\right] + F(z) \tag{D.8}$$

The boundary condition is:

$$\mu_o K_m = -\left.\frac{\partial A_m}{\partial r}\right|_{r=a} \tag{D.9}$$

Once the derivative is taken, the integrals are regular except for the one where m – n = 0. Let $\tau = r/a$. The problem term becomes:

$$\lim_{\tau\to 1}\int_{-\pi}^{\pi}\frac{g_m(\theta')(\tau-\cos(\theta-\theta'))}{\tau^2+1-2\tau\cos(\theta-\theta')}d\theta'$$

Smith argues that as $\tau \to 1$, the integral behaves as if the integrand is $g_m(\theta')[\pi\delta(\theta-\theta')+0.5]$. I have numerically verified this behavior with MATLAB. Using this version of the integrand and the Fourier series version of g_m leads to:

$$\int_{-\pi}^{\pi} g_m(\theta')[\pi\delta(\theta-\theta')+0.5]d\theta' = \pi g_m(\theta) + \pi \tag{D.10}$$

For $m \neq n$ and $t = d(m-n)/a$ define:

$$K_{mnp}(t,\theta,\theta') = \frac{1+t\cos(\theta)-\cos(\theta-\theta')}{t^2+2-2\cos(\theta-\theta')+2t(\cos(\theta)-\cos(\theta'))}\cos(p\theta') \tag{D.11}$$

Putting all this into (D.9) gives the basic equation for the series coefficients.

$$\sum_{p=1}^{q} a_{mp}\cos(p\theta) = \frac{1}{\pi}\int_{-\pi}^{\pi}\sum_{\substack{n=1\\ n\neq m}}^{N}\left[K_{mn0} + \sum_{p=1}^{q} K_{mnp}a_{np}\right]d\theta', \quad 1 \le m \le N \tag{D.12}$$

The integral only applies to the K_{mnp} functions, so we can simplify the notation by defining a general integral form.

$$\begin{aligned} I_{mnp}(t,\theta) &= \frac{1}{\pi}\int_{-\pi}^{\pi} K_{mnp}(t,\theta,\theta')d\theta' \\ &= \frac{1}{\pi}\int_{-\pi}^{\pi}\frac{1+t\cos(\theta)-\cos(\theta-\theta')}{t^2+2-2\cos(\theta-\theta')+2t(\cos(\theta)-\cos(\theta'))}\cos(p\theta')d\theta' \end{aligned} \tag{D.13}$$

With this notation we can simplify and rearrange (D.12) into the standard vector-matrix form.

$$\sum_{n=1}^{m-1}\sum_{p=1}^{q} I_{mnp}a_{np} - \sum_{p=1}^{q}\cos(p\theta)a_{mp} + \sum_{n=m+1}^{N}\sum_{p=1}^{q} I_{mnp}a_{np} = -\sum_{\substack{n=1\\ n\neq m}}^{N} \mathrm{I}_{mn0}, \quad 1 \le m \le N \tag{D.14}$$

The integral can be done numerically quite easily and conveniently, but when run hundreds of times as in loopstick.m, the problem takes a significant amount of time. Closed-form expressions are given in [1] and [2], and I give them here with a slight correction.

$$I_{mnp}(t,\theta) = \frac{D}{\left(1-s^2\right)\left(-s\right)^{p+1}}\left(As^2+Bs+C\right) \tag{D.15}$$

in which

$$\begin{gathered} s = \sqrt{t^2+1+2t\cos(\theta)} \\ A = \cos(\theta-(p-1)\psi) \\ B = 2(1+t\cos(\theta))\cos(p\psi) \\ C = cso(\theta+(p+1)\psi) \\ D = \begin{cases} 1, & p > 0 \\ 2, & p = 0 \end{cases} \end{gathered}$$

$$\psi = \begin{cases} \pi - \tan^{-1}\left(\dfrac{\sin(\theta)}{t+\cos(\theta)}\right), & m > n \\[2ex] -\tan^{-1}\left(\dfrac{\sin(\theta)}{t+\cos(\theta)}\right), & m < n \end{cases}$$

D.1.2 Solution for the Current Coefficients

Equation set (D.14) has Nq unknowns. The coefficients of the unknowns are functions of θ so a simple way to generate enough equations to find the unknowns is to evaluate each of the equations over a number of angles. Smith wrote a least-squared-error integral and then took the partial derivative with respect to each current coefficient to generate enough equations. Rather than implement this method, I chose to go the simpler route of evaluating the N equations at enough angles so that there will be more equations than unknowns. The matrix solution for this situation is a least-squared-error solution. The organization of the equation matrix is as follows. Remember that m and n are the wire counters and p is the space harmonic counter. The elements in each row go from the first wire with its space harmonics from 1 to q (q values), to the second wire with its q harmonics, and so on to the Nth wire with its q harmonics. Now for each m, (D.14) is evaluated at n_θ angles. This makes n_θ rows for each m, making a total of Nn_θ rows. Each row index is $j = i + (m - 1)n_\theta$, $1 \le i \le n_\theta$, and each column index is $k = p + (n - 1)q$. The angles are chosen on the positive half-circle, as using the full circle in the usual stepping manner would produce redundant equations because of the cosine symmetry. I have found that choosing $n_\theta = 2q$ gives good results, comparing against Table I of [2].

Physical symmetry can be used to reduce the number of unknowns. A pair of wires on either side of the group's center and equally distant from it sees the same arrangement of neighboring wires, except that one is the mirror image of the other. $g_3(\theta) = g_{N-2}(\pi - \theta)$, $N > 3$, for example. Since $\cos(p(\pi - \theta)) = (-1)^p\cos(p\theta)$, $a_{3p} = (-1)^p a_{N-2,\,p}$. For N even, one can construct a matrix with half as many columns and half as many rows as one would without symmetry. For N odd, the center wire has no mirror, so $q(N + 1)/2$ is the minimum number of unknowns, and the effects of the center wire have to be treated outside of the main entry-generating loops. I have done this reduced-matrix version for both N even and N odd. The N even case works well, but the N odd case does not produce results as good as I would like. In particular, when there is an odd number of wires, the current on the middle wire should have no odd harmonics because it is influenced equally from both sides. This doesn't quite happen for the minimum-matrix version. It turns out that this even symmetry can be enforced if I use all Nn_θ rows implied by (D.14). So now, instead of generating a reduced matrix element-by-element, I generate the full matrix and then fold it over on itself. The columns for wire N run from $(N-1)q+1$ to $(N-1)q+q$ and they are the mirror for wire 1, columns 1 to q. So column $(N-1)q+1$ is multiplied by -1 and added to column 1, column $(N-1)q+2$ is added to column 2, column $(N-1)q+3$ is subtracted from column3, and so on. If the full matrix is M, Nn_θ by Nq, and the reduced-size matrix is Mr, Nn_θ by $Nq/2$ for N even, the process in MATLAB reads:

```
for n=1:N/2
    for p=1:q
        Mr(:,p+(n-1)*q)=M(:,p+(n-1)*q)+M(:,(N-n)*q+p)*(-1)^p;
    end
end
```

It requires about twice as much time to fill this matrix as to fill the minimum-size matrix but the inversion time is about the same.

Figure D.2 shows the current distributions on selected wires in a 41-wire group. Wire 21 is the center wire, and it has a symmetrical distribution with maxima on the top and bottom surfaces. Note that each figure is to a different scale. The other wires have top/bottom symmetry, but not left/right symmetry. This is because there are more currents on their right pushing their current left than otherwise. Even with wire 15 there is negative current—that is, current flowing opposite to the average direction. At wire 1, the distortion is at its maximum, with a normalized maximum positive flow of about 5.43 A/m and a normalized negative maximum of about –10.1 A/m. Remember when contemplating these curves that the average has to be 1.

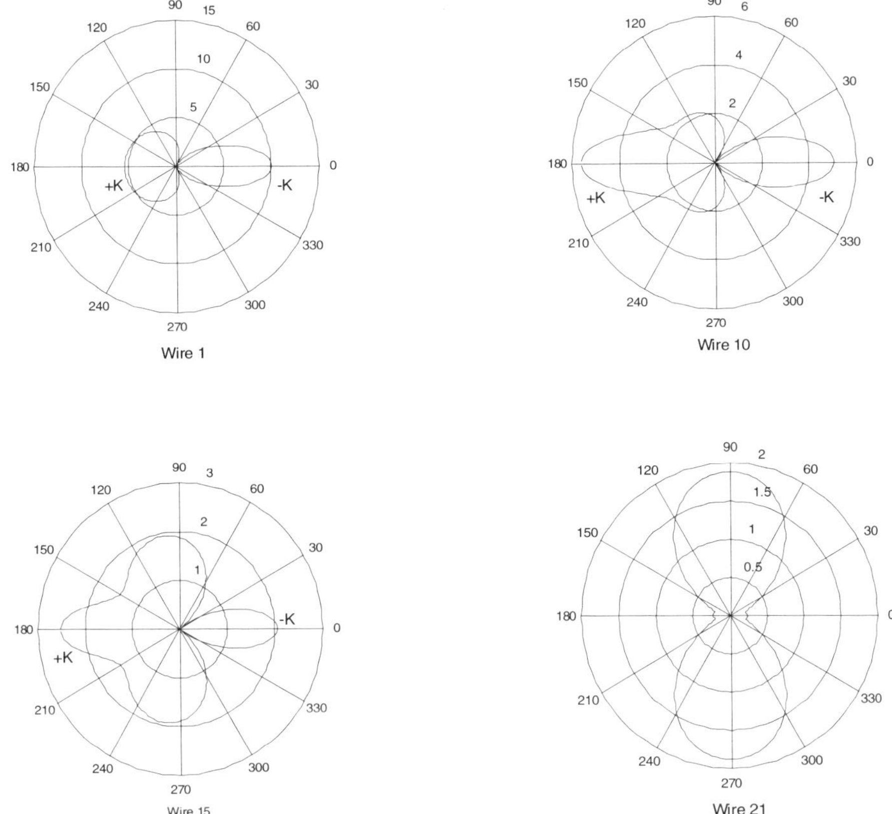

Figure D.2: Current distributions on selected wires in a 41-wire group. The plots show the absolute value of the current density. Except for wire 21, the right-hand lobe in each plot is negative current, its flow is opposite to the reference direction. Wire 21 is the center wire, the others are on the left of the group. d/a = 2.2. The normalized excess resistance is R_p/R_o = 8.3; see section D.2.

D.2 Power and Resistance

The analysis for current distribution assumes that the skin effect is such that the current is forced into a thin layer under the surface of the wire. For power loss, we can say that the loss is a small perturbation on the lossless distribution. The power loss per unit surface area is the square of the current density divided by the skin depth and conductivity. For wire m,

$$p_m(\theta) = \frac{K_m^2(\theta)}{\sigma\delta_s} = K_m^2(\theta)\sqrt{\frac{\omega\mu_o}{2\sigma}} = \sqrt{\frac{\omega\mu_o}{2\sigma}}\,\frac{I^2}{4\pi^2 a^2}\left(1+\sum_{p=1}^{q} a_{mp}\cos(p\theta)\right)^2 \qquad \textbf{(D.16)}$$

where I is assumed to be rms current. The total power per unit length is the integral around the wire of $p_m(\theta)$,

$$P_m = \int_{-\pi}^{\pi} p_m(\theta) a d\theta = \sqrt{\frac{\omega\mu_o}{2\sigma}} \frac{I^2}{4\pi^2 a} \int_{-\pi}^{\pi} \left(1 + \sum_{p=1}^{q} a_{mp} \cos(p\theta)\right)^2 d\theta$$
$$= \sqrt{\frac{\omega\mu_o}{2\sigma}} \frac{I^2}{4\pi^2 a} \left(2\pi + \pi \sum_{p=1}^{q} a_{mp}^2\right) = \sqrt{\frac{\omega\mu_o}{2\sigma}} \frac{I^2}{2\pi a} \left(1 + \frac{1}{2} \sum_{p=1}^{q} a_{mp}^2\right) \quad \text{(D.17)}$$

Since the series is orthogonal, the integration takes out all the harmonic cross-products, leaving a sum-of-amplitude-squares. In the last expression, the term outside the parentheses is the ordinary skin-effect resistance loss per unit length for a single wire, I^2R_o. Therefore, the summation inside the parentheses is the normalized excess resistance due to proximity,

$$\left.\frac{R_p}{R_o}\right|_m = \frac{1}{2} \sum_{p=1}^{q} a_{mp}^2 \quad \text{(D.18)}$$

For N wires, the average excess ratio is:

$$\frac{R_p}{R_o} = \frac{1}{2N} \sum_{m=1}^{N} \sum_{p=1}^{q} a_{mp}^2 \quad \text{(D.19)}$$

When using a reduced-matrix method to solve the problem, it is best to reconstruct the rest of the coefficients before using (D.19).

Smith[2] describes an elegant experiment in which he measured the current distributions on closely-spaced parallel cylinders. The data points lie nicely on the calculated curves.

References

[1] G. S. Smith, "Proximity Effect for Systems of Parallel Conductors and Electrically Small Loop Antennas," Engr. Tech. Rep. 624, Gordon McKay Library, Harvard University, Dec. 1971.

[2] G. S. Smith, "Proximity Effect in Systems of Parallel Conductors," J. App. Phys. vol. 43, no. 5, pp. 2196–2203, May, 1972.

APPENDIX E

What Every EE Student Should Know About Mathematics by the Senior Year

E.1 What Is Mathematics to an Engineer?

Mathematics is an extension to natural language by which we express abstractions, relations, and facts useful in practice. We use its concepts and notation to construct models of the physical world and apply logic to these models to discover their performance properties, just as logic should be applied to word descriptions of situations in other human activities. To communicate in engineering, people on both sides of the conversation must have a basic facility in the language. A person lacking a basic knowledge of the relevant language will spend too much time trying to translate (or give up) and barely or never get the real information intended.

The material in the following sections is a subset of what you should know about mathematics. It does represent a collection of things that you should know as well as the meaning of "blue" and operations you can do as simply as boiling water.

E.2 The Process Is as Important as the Result

How is the basic language facility acquired? Of course, one takes a number of courses to get started. Then, just like going to Germany to learn German, you take engineering classes in which certain parts of the total area of mathematics are used over and over again. You become proficient by working through problem developments in your texts and in classes. Try not to take anything for granted.

Derivations in engineering are not like proofs in mathematics because the concerns of the mathematicians and the engineers are generally different. Mathematicians are concerned about the existence and uniqueness of solutions. Engineers generally feel they have a physical basis for knowing whether or not a solution exists. About the only place where mathematicians and engineers have the same concern is in the convergence of iterative methods. While you may have felt that the ε - δ proofs in your math courses were not too important since they lacked connection to physical reality, the derivations in your engineering classes usually proceed by setting up a physical model, using some basic mathematics, some results from earlier physical models, and working (or forcing) the model relationships to get the desired information. This *process* is central to engineering activity. Only by practicing this process will you learn a number of facts and formulas which are analogous to idioms and essential to real engineering conversation.

E.3 Facts and Idioms

This section lists special numbers, trig identities, and some approximations that occur so frequently that you should know them by heart and recognize them whenever needed. It is set up as a self-test, with the answers on following pages.

E.3.1 Special Numbers

(a) A calculator will give you the values of these numbers to many decimal places, but it is often useful for checking to know them to a few significant figures yourself.

$$\pi \quad e \quad \sqrt{2} \quad 1/\sqrt{2} \quad \sqrt{3} \quad \sqrt{5} \quad \sqrt{10}$$

(b) Trig functions are periodic, with their basic argument range usually taken to be from $-\pi$ to $+\pi$ or 0 to 2π. You should know the sine, cosine, and tangent values for multiples of $\pi/4$ and $\pi/6$ in these ranges, in terms of the special numbers for their basic triangle side lengths. For example, $1/\sqrt{2}$ or $\sqrt{3/2}$.

(c) You should be equally at home with polar, rectangular, and exponential representations of complex numbers. Some special values related to the trig functions given above are:

$$1+j \quad 1-j \qquad -1+j \quad -1-j$$

$$\sqrt{3}+j \quad -1+j\sqrt{3} \qquad 0+j \quad -1+j0$$

List the polar and exponential forms.

E.3.2 Identities and Formulas

(a) $\cos(a)\cos(b)$ in terms of sum and difference angles.

(b) $\sin(a)\cos(b)$ in terms of sum and difference angles.

(c) $\sin(a)\sin(b)$ in terms of sum and difference angles.

(d) $\sin(a + b)$.

(e) $\cos(a + b)$.

(f) $\sin^2(a) + \cos^2(a)$.

(g) Euler's Identity.

(h) Roots of $x^2 + bx + c = 0$ and $ax^2 + bx + c = 0$.

(i) $(a + b)(a - b)$.

(j) $(a + jb)(a - ji)$.

(k) $(a + jb)(c + jd)$.

(l) $(a + jb)/(c + jd)$.

E.3.3 Approximations

Many of our models of physical problems involve idealizations in which some quantity is small compared to another. This allows us to approximate nonlinear behavior with simple expressions, frequently with either a constant or a linear one. Write appropriate one- or two-term approximations for the following, assuming $x << 1$.

$\sin(x)$ $\cos(x)$ $\tan(x)$ e^x $\ln(1 + x)$ $(1 + x)^p$ $1/(1 + x)$

E.4 Integrals and Derivatives

You should know the derivatives and integrals of the trig, log, and exponential functions mentioned above, and the power function. For differentiation, the chain, product, and quotient rules should feel intuitive. For integration, the most common strategies are decomposition into known forms, and integration by parts. Of course, you should own a handbook that includes integral tables because there are many integrals needed occasionally, but not so often they'll stick in the mind.

Quite frequently, you need to find the limiting value of an apparently indeterminate quantity. This quantity can usually be set up as a ratio, 0/0 or ∞/∞, and l'Hospital's Rule applied. Alternatively, the numerator and denominator can be converted into functions of x, x a small value, and the approximations mentioned above will serve to show the limit as x goes to zero.

E.5 Radians or Degrees?

As a measure of angle, the degree is deeply imbedded in our culture, but it has no particular mathematical or physical utility. The radian is defined as arc length on a circle of unit radius, which is also the defining circle for the trigonometric functions and $e^{j\theta}$. When integrals or derivatives with respect to angle are done using radian measure, the function changes directly because the arc length changes, so no unit conversion is needed. If you haven't seen this before, you might try finding the derivative of the cosine by geometric construction.

Many of our results are derived using calculus on such trig functions. While it is possible to work in degrees by keeping an eye out for πs that will need to be replaced by 180, converting ω to degrees/second, and so on, it really isn't worth the trouble. The best policy is to use radians in all calculations, convert initial data to radians, and make sure your calculator or software is in radian mode. A lot of otherwise good work has been turned to hash just by the mode error.

E.6 Matrix Notation and Operations

Matrix notation is first of all a shorthand for the collection of sets of equations. For most of our problems, these equations describe relations among input signals, internal system or circuit signals, and output signals. Each signal set is usually collected into a column vector, and the coefficients that relate them are collected into rectangular matrices. When you write circuit or system scalar equations, you should do so with the variable labels having the same order of occurrence in each equation, so that you can directly transfer the coefficients to matrices. When you read a matrix equation, always think about the scalar equation each row of coefficients represents. This will help you to hang on to the physical meaning of the notation.

Of course, if you're going to use a shorthand, you have to have rules so that the results come out the same as if you had dealt with the scalar equations directly.

This is the basis of the multiplication and addition rules, and those governing the manipulation of transpose and inverse. You should know the meanings of, and how to calculate analytically, the following: trace, transpose, minor, determinant, cofactor, adjoint, and inverse.

E.7 Answers for Section E.3

E.3.1

(a) 3.14 2.718 1.414 0.707 1.732 2.236 3.16

(b) $\theta = k\pi/4$

k	0	1	2	3	4	5	6	7
sinθ	0	$\frac{1}{\sqrt{2}}$	1	$\frac{1}{\sqrt{2}}$	0	$-\frac{1}{\sqrt{2}}$	−1	$-\frac{1}{\sqrt{2}}$
cosθ	1	$\frac{1}{\sqrt{2}}$	0	$-\frac{1}{\sqrt{2}}$	−1	$-\frac{1}{\sqrt{2}}$	0	$\frac{1}{\sqrt{2}}$
tanθ	0	1	±∞	−1	0	1	±∞	−1

$\theta = k\pi/6$

k	0	1	2	3	4	5
sinθ	0	$\frac{1}{2}$	$\frac{\sqrt{3}}{2}$	1	$\frac{\sqrt{3}}{2}$	$\frac{1}{2}$
cosθ	1	$\frac{\sqrt{3}}{2}$	$\frac{1}{2}$	0	$-\frac{1}{2}$	$-\frac{\sqrt{3}}{2}$
tanθ	0	$\frac{1}{\sqrt{3}}$	$\sqrt{3}$	±∞	$-\sqrt{3}$	$-\frac{1}{\sqrt{3}}$

k	6	7	8	9	10	11
sinθ	0	$-\frac{1}{2}$	$-\frac{\sqrt{3}}{2}$	−1	$-\frac{\sqrt{3}}{2}$	$-\frac{1}{2}$
cosθ	−1	$-\frac{\sqrt{3}}{2}$	$-\frac{1}{2}$	0	$\frac{1}{2}$	$\frac{\sqrt{3}}{2}$
tanθ	0	$\frac{1}{\sqrt{3}}$	$\sqrt{3}$	±∞	$-\sqrt{3}$	$-\frac{1}{\sqrt{3}}$

(c) $\sqrt{2}\angle\frac{\pi}{4} = \sqrt{2}e^{j\frac{\pi}{4}}$ $\quad\sqrt{2}\angle-\frac{\pi}{4} = \sqrt{2}e^{-j\frac{\pi}{4}}$ $\quad\sqrt{2}\angle\frac{3\pi}{4} = \sqrt{2}e^{j\frac{3\pi}{4}}$ $\quad\sqrt{2}\angle\frac{-3\pi}{4} = \sqrt{2}e^{-j\frac{3\pi}{4}}$

$2\angle\frac{\pi}{6} = 2e^{j\frac{\pi}{6}}$ $\quad 2\angle\frac{2\pi}{3} = 2e^{j\frac{2\pi}{3}}$ $\quad j = 1\angle\frac{\pi}{2} = e^{j\frac{\pi}{2}}$ $\quad -1 = 1\angle\pm\pi = e^{\pm j\pi}$

E.3.2

(a) $\frac{1}{2}\left[\cos(a+b)+\cos(a-b)\right]$

(b) $\frac{1}{2}\left[\sin(a+b)+\sin(a-b)\right]$

(c) $\frac{1}{2}\left[\cos(a-b)-\cos(a+b)\right]$

(d) $\sin(a)\cos(b) + \cos(a)\sin(b)$

(e) $\cos(a)\cos(b) - \sin(a)\sin(b)$

(f) 1

(g) $\cos(\theta)+j\sin(\theta) = e^{j\theta}$

(h) $\frac{-b}{2}\pm\sqrt{\frac{b^2}{4}-c}, \quad \frac{-b\pm\sqrt{b^2-4ac}}{2a}$

(i) $a^2 - b^2$

(j) $a^2 + b^2$

(k) $ac - bd + j(ad + bc)$

(l) $\frac{ac+bd+j(bc-ad)}{c^2+d^2}$

E.3.3

$x \quad 1-\frac{x^2}{2} \quad x \quad 1+x \quad x \quad 1+px \quad 1-x$

Index

Symbols

A

B

N

O

P

Q